GRAFTON ELLIOT SMITH, EGYPTOLOGY & the DIFFUSION of CULTURE

GRAFTON ELLIOT SMITH, EGYPTOLOGY & the DIFFUSION of CULTURE

A Biographical Perspective

PAUL CROOK

SUSSEX
ACADEMIC
PRESS
Brighton • Portland • Toronto

2 4 6 8 10 9 7 5 3 1

First published in 2012 by
SUSSEX ACADEMIC PRESS
PO Box 139
Eastbourne BN24 9BP

and in the United States of America by
SUSSEX ACADEMIC PRESS
920 NE 58th Ave Suite 300
Portland, Oregon 97213-3786

and in Canada by
SUSSEX ACADEMIC PRESS (CANADA)
90 Arnold Avenue, Thornhill, Ontario L4J 1B5

British Library Cataloguing in Publication Data
A CIP catalogue record for this book is available from the British Library.

Library of Congress Cataloging-in-Publication Data
Crook, D. P. (David Paul)
Grafton Elliot Smith, Egyptology and the diffusion of culture : a biographical perspective / Paul Crook.
p. cm.
Includes bibliographical references and index.
ISBN 978-1-84519-481-9 (pbk. : alk. paper)
1. Smith, Grafton Elliot, Sir, 1871–1937. 2. Anthropologists—Great Britain—Biography. 3. Anthropologists—Australia—Biography. 4. Egyptologists—Great Britain—Biography. 5. Egyptologists—Australia—Biography. 6. Culture diffusion. 7. Ethnology—Methodology. 8. Physical anthropology. I. Title.
GN21.S63C76 2012
301.092—dc23
[B]

2011016463

Typeset and designed by Sussex Academic Press, Brighton & Eastbourne.
Printed by TJ International, Padstow, Cornwall.
This book is printed on acid-free paper.

Contents

Preface

Grafton Elliot Smith (1871–1937) is a great forgotten Australian. He was one of the world's pioneering anatomists, an authority on human evolution, and a renowned, if controversial, amateur archaeologist/anthropologist. He wrote numerous scholarly and popular works, founded a leading edge medical and social science school at University College of the University of London, and was made a fellow of the Royal Society and a knight of the realm. Yet today his name is virtually unknown in the land of his birth, partly no doubt because he lived most of his life in Britain – although he travelled widely and often revisited his homeland, not least significantly when he was hugely instrumental in setting up anthropology as an academic discipline in Australia.

Elliot Smith is worth remembering, and I hope that this little book will help in that respect. It is not a biography as such, but rather a history of the man and his ideas put in the context of his life and times, with the major focus on his much contested theory of the diffusion of culture, which put Egypt as the fountain-head of human civilization, the centre from which major elements of civilization were spread by the migration of peoples and ideas. I want to revisit his writings, robust and challenging, but always scientific in their methodology; to see them in the light of contemporary events – such as the exciting archaeological discoveries of the early twentieth century and the catastrophic First World War – and to look at the way people reacted to his theories. I do not wish to put forward any hagiographical or total defence of his position; but at the same time I believe that some redressing of the intellectual balance is required. His diffusionist model may not have become – as it at one stage seemed to promise to become – the ruling paradigm in anthropology, but nor has it been conclusively refuted, despite being much ridiculed in some academic circles. Elliot Smith didn't win his debate – although he contended to the end against such heavyweights as Bronislaw Malinowski that it was winnable. Rather than being refuted by systematic research, it is contendable that the discourses of anthropology and archaeology simply moved on to other issues and embraced other methodologies. Many central ideas raised by Elliot Smith and co-diffusionists such as W. J. Perry were essentially side-stepped and never really subjected to sustained scrutiny. Elliot Smith would have been the first to welcome such scrutiny. Scientific progress, he always said, was a matter of offering hypotheses and testing them rigorously, and he always professed himself more than willing to accept conclusions that contradicted his own suggestions. My overall message is that Elliot Smith's prodigious labours and fertile ideas – so long unfairly caricatured and stereotyped in the ethnological literature – deserve considered reassessment.

Whilst I was researching Elliot Smith's life it was brought home forcefully to me many times that the people I was dealing with – colourful figures such as W.H.R. Rivers, Will Perry, Bronislaw Malinowski, John Linton Myres and others – were not reasoning machines or purely cerebral creatures, but were living, breathing humans with the emotions, feelings and agendas of the human condition, products of their time and cultures. Certainly this was true of Elliot Smith. He varied from being kind, charming and urbane to being cantankerous and combative (especially in print). He may even have been a trifle autistic, given his capacity for fierce concentration and indifference to personal circumstances. For example, he was oblivious to the flies and heat of the Egyptian desert when he was examining thousands of mummies that were being uncovered in the famous excavations of the 1900s. As for being a product of his culture, the fact that he was a colonial Australian may not have been entirely irrelevant to the fact that he was inclined to rebelliousness against scientific orthodoxies and often lacking in the usual respect given to ranking thinkers. The irony of it all was that he himself was to become a mandarin in the very scientific establishment that he appeared to disdain.

I am an historian and do not attempt in this book to make technical judgments on the archaeological and anthropological issues at stake. However I would like to acknowledge the welcome advice given by my archaeologist wife Ann, whose loving support and encouragement sustained me throughout this project. Other archaeological friends also kindly offered help and guidance. Most of them were totally unaware of Elliot Smith's writings, which had disappeared from academic reading lists after the Malinowskian revolution. I am deeply thankful also to Margaret Higgs who over many years has provided me with invaluable research assistance and editorial advice. I owe a debt also to the libraries and institutions that hold materials relevant to Elliot Smith. Holdings of his papers are listed in the Select Bibliography. Many thanks to the staff of the Inter-Library Loan section of the University of Queensland Library for obtaining masses of material for me, much of it rare and difficult to access.

DAVID PAUL CROOK
Brisbane

CHAPTER

1

Early Days: Sydney, Cairo, Manchester (1871–1915)

Grafton Elliot Smith [hereafter ES] was born on 15 August 1871 in the country town of Grafton in northern New South Wales in Australia, after which he was named.[1] He seems to have acquired the surname Elliot Smith some time after he arrived in Britain more than twenty-five years later.[2] The double barreled name (without the hyphen), with its upper-class connotations, presumably gave him extra cachet in the British class structure. No-one seems to have called him Grafton, at least in any of the correspondence I have seen. What his wife Kate called him is an interesting question.

His father was Stephen Sheldrick Smith (*c.*1840–1929), a country school master.[3] Stephen Smith was a born and bred Londoner, raised in the Piccadilly Circus precinct. As ES tells us about his father in an autobiographical fragment (unfortunately one of the few personal memoirs that has survived): "As a child he obtained employment in the shop of Winsor and Newton, who sold artists' materials, concerning which he always retained exceptional knowledge. When the Working Men's College was started in Queen Square, he enrolled as a student under [John] Ruskin and attained sufficient knowledge to obtain the position as a teacher in the Educational Department at Sydney, when his father emigrated thither with his family – an enterprise which took six months in a sailing ship. It was from my father that I acquired an interest in drawing with pencil and pen, an occupation to which I devoted much attention in childhood".[4] ES was to illustrate his own works in the future.

The Elliot in Elliot Smith came more immediately from ES's grandfather Elliot Smith (*c.*1817–1880),[5] who emigrated to Melbourne with his wife Eliza, three daughters and a son (Stephen Sheldrick) in 1857.[6] The name came ultimately from a revered ancestor who was mayor of Cambridge in 1851, 1859 and 1867. ES recalled: "When I became a student at Cambridge, I was interested to see, on the cast-iron milemarks in the roads to Cherry Hinton, the name ELLIOT SMITH".[7]

Stephen Sheldrick Smith married Sydney-born Mary Jane Evans, of Welsh descent, in 1864.[8] There seem to have been good intellectual genes on both sides of the family. The eldest son Stephen Henry (1865–1943), born in Wollombi in rural New South Wales, became director of education in the state. ES's younger brother Stewart Arthur became a demonstrator, and later acting professor, in anatomy at the University of Sydney.[9]

ES's father was an enthusiast for education and seems to have ignited a scientific interest in his son, who was soon observing and collecting specimens of fauna and flora in the nearby bush. ES recalls how when he was ten his older brother (called Henry) on one occasion returned from study at the teacher's training college in Sydney, bringing a little book on elementary physiology "which opened a new vision to me and I became fascinated by this new field of interest. Ever afterwards, when on the vacation visits to the seaside I found the carcase of a dead shark on the beach, I proceeded to dissect it with a penknife, and became specially intrigued by the brain which seemed to me to be a veritable collection of puzzling tricks".[10] Soon after this his father was transferred to Sydney. They lived in a house at Darlington. Here there was an interesting connection with the famous Thomas Henry Huxley, Charles Darwin's staunchest supporter in battles over the theory of evolution that then raged in England, and who became a hero to ES. It was from Huxley that he imbibed an enthusiasm for making science accessible to a general audience. A short distance from the Smith's house was a house that had been described and sketched by Huxley in his well-known *Voyage of H.M.S. Rattlesnake*. There lived Henrietta Heathorn, whom Huxley met in 1847 and later married. Serendipitously, the Huxleys came to live in a terrace house in Regent's Park, London, where ES later resided (where in fact he penned his fragmentary memoir, shortly before his death).

ES continued his education at the Darlington public (government) school, where his father was principal, then at Sydney Boys' High School. He recalled how each day he walked from Newtown into the city: "This way led me past the School of Technology and I was attracted by a notice to the effect that Professor Anderson Stuart, the Dean of the Faculty of Medicine in the University, was giving a course of instruction in physiology in the evenings. I promptly signed on as a member of this class, anxious to satisfy the curiosity which had been aroused. The class more than satisfied my hopes". He was thrilled by the information and lucidity of Huxley's *Elementary Lessons in Physiology* and "with my appetite thoroughly stirred, I began to frequent the Public Library to discover what other thrills scientific books held . . . Eventually the day arrived when

Professor Anderson Stuart invited his class to visit the Medical School, and showed us, and allowed us to handle, human brains. He called our attention to the convolutions and Aeby's figures of them and told us that no one knew them all. I remember silently framing the vow that I should be the one exception to this statement".[11] He was admitted to the medical school in Sydney University in 1888, but not without difficulties. He failed earlier subjects due to spelling fallibilities, but persisted – against his father's desire for him to enter insurance. He won entry only by winning medals, some of which were in subjects he studied that were not offered at school. He was allowed a trial year and eventually won a bursary from the university after obtaining a prize for physics and natural history. His subsequent career was little less than spectacular.

The medical school at Sydney at that time had brought together a highly talented staff, many of them young men who went on to distinguished careers in the UK. The department of physiology had Anderson Stuart, Almoth Wright (who became head of pathology at St Mary's Hospital in London) and Charles J. Martin (later director of the Lister Institute in London); Alexander MacCormick became head of the surgery department; there was Carmack Wilkinson in pathology; while in science generally there were people like Richard Threlfall in physics, Edgeworth David, W. A. Haswell and J. P. Hill in geology. Wright, Martin and Threlfall were later knighted.[12] ES worked under J. T. Wilson, who saw a great future for his young protégé. His years in Sydney gave ES a lifelong respect for cross-fertilization across a range of disciplines, respect for what Robert Oppenheimer later described as the cooperative and interrelated character of scientific achievement.

ES graduated as a bachelor of medicine (MB ChM) in 1892. He spent the next years in clinical practice and in anatomical research for a doctorate. After a year as a house-surgeon he was invited by Wilson to become his demonstrator in anatomy. ES craved to do original research in neurology, subscribing to the journal *Brain* from his first year and becoming fascinated with recent work on the thalamus and sensory paths. Charles Martin suggested that he work on the spino-thalamic tract in Australian marsupials, which were plentiful (both live and preserved) and on which little or no work had been done. (Traditionally researchers worked on cats.) ES collected bandicoots and other marsupials. Within a year he had published in the *Proceedings of the Linnean Society of N.S.W.*, a ground-breaking paper on marsupial neurology (1894). It was to become a classic paper in the literature. During his two years as a demonstrator he published "some dozen papers on neurological subjects, including: the relations of the olfactory bulb and hippocampus; the morphology of the cerebellum; Jacobsen's organ in *Ornithorhynchus*; the cerebrum of *Notoryctes*; the 'true limbic lobes, etc.' . . . ".[13] His doctoral thesis on the anatomy and histology of the brain of non-placental mammals won him a gold medal and he was awarded his doctorate in medicine (MD) in 1895. It was time to look at wider horizons.

The British Empire was still at the height of its power in the 1890s. This was despite some ominous signs. Britain's industrial supremacy in the world was being challenged by rising nations such as newly-united Germany and the United States, just as its colonial and naval supremacy was being threatened by imperial rivals. With the rise of a militarized Germany, there was great power rivalry and an alarming arms race that was ultimately to result in the Great War of 1914–18. White European peoples also felt menaced by a population explosion in Asia, a "yellow peril" that threatened to swamp them. Modernism and avant-garde cultural movements gave many a sense that they were living in a crisis time of alienation and decadence.

However there was an essential optimism about the empire in the robust settler societies of the "Anglo-world", Australia, New Zealand, Canada, South Africa, Charles Dilke's *Greater Britain*.[14] Britain – or England to most – was still the Mother Country, and the sun never set on the British Empire, marked red, or pink, on all school maps. Well-off Australians did their almost obligatory "Grand Tour" of the Mother Country. As for education, there was a well trodden track from Sydney, Melbourne, Brisbane, Adelaide and Perth to Britain that was taken by graduates wanting to do postgraduate research in their chosen fields. Very little postgraduate research was done in Australia at the time. The universities of London, Oxford and Cambridge were prime destinations for higher education, sometimes Edinburgh or provincial universities such as Manchester or Birmingham.

In April 1896 ES set out for London on the R. M. S. *Himalaya*, having won a two-year James King traveling scholarship from Sydney University. He was "full of enthusiasm but with no very definite plans and armed with letters of introduction from his Sydney friends to the leading anatomists and biologists in Britain".[15] He followed these contacts up with a vengeance, in six months meeting many of Britain's leading anatomists, doctors and zoologists. He first looked up T. H. Huxley's successor as professor of zoology at the Royal College of Science in London, George Bond Howes, who was to mentor him in many ways, sponsoring a series of ES's anatomical papers. His name was becoming known in medical circles, but forging an academic career in Britain was by no means a foregone conclusion for an Australian "outsider". The country boy from Grafton still had some rough edges: "Elliot Smith was thought by many, at least in his earlier years, to be reserved in manner. The easy sociability flowered later."[16] Yet at his centre there was a tough confidence in his own abilities.

He wrote "innumerable" letters to scholars: "What spare time I have been able to get has been spent in the reading room of the British Museum and Royal College of Surgeons where I have been revelling in a lot of literature which it was impossible to get in Sydney. I have also 'been through' the brain material at the R.C.S.: but it is in a fearfully lacerated condition".[17] He met leading anatomists at a meeting of the Anatomical Society at Oxford in July 1896, and visited various anatomy schools. He considered a number of medical posts but finally settled on Cambridge, after meeting Alexander Macalister there in late July. Macalister was professor of anatomy at Cambridge, and he suggested that ES enter there as a Research Student, a position only recently established at Oxford and Cambridge. Macalister was to have a formative influence upon ES: "Macalister is a very busy man and devotes all his time that is not absorbed in the study of Early Church History and Oriental philology to the personal superintendence of the work in the Anatomical school. He seems, however, to be intensely devoted to Anthropology of, I am sorry to say, the bone-measuring variety".[18] Significantly Macalister had an interest in Egypt.

ES enrolled as a Research Student of St. John's College at the opening of the academic year, 1 October 1896, "with adequate facilities for anatomical research in the midst of an active band of workers in various branches of biological science". In 1897, in what was virtually an international congress of anatomists at Dublin, "he read a remarkable paper on the relation of the fornix to the margin of the cerebral cortex" and a string of impressive papers followed.[19] In 1898 he became the first biologist to take the recently created Research B.A. degree. This was the equivalent of the later Ph.D. It was only in 1895 that Cambridge had allowed qualified graduates of other universities to do advanced research. They were allowed to gain the research B.A. after two years by submitting a thesis based on original research. Among the young graduates from the Dominions who took advantage of this opportunity was the New Zealander Ernest Rutherford. He and ES became good friends (and were later to be reunited in Manchester). Rutherford left a description of ES at age 26: "He was shy and taciturn at first with strangers, while his drooping moustache gave him an appearance almost of melancholy. This soon vanished when he talked with friends on matters in which he was interested, when he became lively and humorous and the best of company. His outward appearance changed markedly in middle age, when he was clean-shaven, rubicund, and a ready and fluent speaker, looking, to my mind, rather like a distinguished and jovial bishop".[20]

In the meantime ES was characteristically busy, working in the physiological laboratory, attending meetings of scholarly bodies such as the Physiological Society in London, even applying for the chair of anatomy in Cardiff (although he confessed to being too junior for this or other posts that were suggested to him by Macalister), and in July of 1897 voyaged to Canada for the British Association meeting being held there. As usual by now he was in distinguished company. Although he was a research appointment, he ensured that he did some teaching at Cambridge "for the mere glory of the thing" (and also no doubt as insurance for his future prospects). He also edited the section on the central nervous system for the *Journal of Anatomy and Physiology*.[21] In the spring of 1898 the British Medical Association granted him a scholarship but the deal fell through as it entailed him shifting to London. Thus, his Sydney scholarship having run out, "I set out at the beginning of October to earn my bread and cheese independently and for three months endured a life of abject slavery. I used to demonstrate for the whole of every morning, alternately in anatomy and physiology, I gave lectures on the brain twice a week and coached every afternoon and evening. You can imagine the agony of such a life to my lazy temperament . . . ".[22] Presumably this was said tongue in cheek. Between 1897 and 1900 he published eight important papers dealing with cerebral morphology.[23] He gained security when in November 1899 he was elected a Fellow of St. John's College: "His election to the Fellowship was especially a matter for congratulation, because he was a student of only three years' residence (five years being the minimum which custom had fixed for eligibility), and at that time, moreover, the University authorities did not always look kindly upon 'Colonials'."[24]

In 1899, at the invitation of the Council of the Royal College of Surgeons, he undertook to prepare a descriptive and illustrated catalogue of the unrivalled collection of brains preserved in their Museum (the same collection he had described as in "fearfully lacerated condition"): "Published in 1902 this is still a standard work on the mammalian, and especially the primate, brain".[25] For this enormous task he received the princely sum of £100.

ES now regarded himself as "happily placed" in Cambridge, even spurning approaches to go to America. Yet in mid-1900 he accepted an offer, passed on to him by Macalister, to become the first professor of anatomy in the new government medical school in Cairo, headed by the Englishman Henry Keatinge. Macalister had visited Egypt the previous winter and had been shown over the school by Keatinge. ES met Keatinge in London in July 1900 and, satisfied by the prospects for research, accepted the challenge. As Macalister had urged upon him: "There is magnificent material for anatomical work there, all the bodies dissected there are natives of one African tribe or another and they provide all the requisites for work there, and whatever will be required to make a Museum . . . The English officials are an exceedingly pleasant set".[26]

This dramatic life change probably accelerated his decision to marry Kathleen (Kate) Emily Macredie, a cousin of Alice Macredie, his brother Henry's wife. He was 29 and she was 26, living in Bloomsbury. ES had met Kate on a trip to Ireland. They married at Chelsea on 22 September 1900, and soon afterwards departed on a honeymoon voyage to Cairo, which they reached in early October. By all accounts the marriage was a happy one (although as Norman Mailer once remarked who knows what a marriage is really like). The union produced three sons (Arthur, born in Cairo, 3 June 1901, Latimer, born in Cairo, 24 September 1903, and Stephen, born in Manchester, 12 December 1910). ES was soon absorbed in the task of organizing the new medical school, a valuable apprenticeship for his later career as head of anatomy in University College, London.

It was a time of growing interest in Egyptology and Cairo was becoming a tourist mecca for wealthy Europeans and Americans. At first he resisted the temptation "to dabble in Egyptology".[27] However in his work he became inundated by mummified human remains being unearthed by archaeologists as a result of the enlargement of the Aswan dam then taking place in Nubia. The teams included the Hearst expedition of the University of California under the distinguished archaeologist George A. Reisner, whose excavations ranged from the Early Predynastic to the Middle Empire. ES began officially cooperating with the official department of Egyptian antiquities.

It was about this time that the psychologist-anthropologist W. H. R. Rivers appeared upon the scene. Rivers was to have a seminal influence upon ES. In fact it may be said that Rivers set ES upon his diffusionist path. As time went on, the two men developed a strong partnership and collaboration, with ES taking a progressively more forceful role and in his own turn influencing Rivers markedly (with some Rivers biographers decrying this as a fatefully wrong step in Rivers's career). It is misleading to see (as some have) either man as a puppet or puppeteer. It was rather a case of two independent minds working empathetically together.

William Halse Rivers Rivers (1864–1922) was a physician who turned to psychology after studying mental health problems, writing papers on hysteria and neurasthenia. In 1897 he was appointed the first officially recognized lecturer in experimental psychology at Cambridge, after studying the subject in Germany and under his mentor, the famous neurologist Hughlings Jackson. Partly by happenstance, but also in order to apply his rigorous experimental methods to ethnology, Rivers was prevailed upon to join A. C. Haddon's famous Cambridge Anthropological Expedition to the Torres Straits in 1898. There he developed methodological techniques in areas such as genealogy and kinship, plus a range of other innovations that were to make him a highly influential figure in early twentieth-century ethnology. Around 1900 he was working on the psychology of the special senses, especially vision, which he had studied in Torres Strait Islanders.

ES later reminisced about Rivers:

> In the winter of 1900–1901, he went out to Egypt and at the archaeological camp of Dr Randall-Maciver and the late Mr Anthony Wilkins at El Amrah near Abydos in Upper Egypt, he investigated the colour vision of the workmen engaged in their investigations. Incidentally and quite unwittingly, Dr Rivers was responsible for drawing me into anthropology. At his suggestion I visited El Amrah to study what until then was the unknown phenomenon of the natural preservation of the brain in the crania of predynastic Egyptians. So many other problems in anatomy presented themselves for solution that before long I had definitely committed to the study of the anthropology of Egypt. In the following year, Dr Rivers went out to India to embark upon his first independent research in pure ethnology – the study of the Todas.[28]

As a cranial anatomist ES was fascinated by the "desiccated brains" that were found in Egyptian skulls of all periods. They had been remarkably preserved in the hot sands. He was soon preoccupied with analysis of their structure, and published his findings in England.[29] This was work that had never really been done systematically before. As an experienced doctor ES was temperamentally suited to dissecting tasks that seemed gruesome to others. It was at this time that he inspected remains at excavation sites, and he continued to visit sites over the next years. He did extensive field work in particular during the Nubian survey of 1907, when he measured and medically examined thousands of remains in burial sites. This did not make him a qualified archaeologist, but he was a great deal more than the "armchair" ethnologists of the time. Their era was rapidly passing. The Torres Strait expedition of 1898 had set an example of rigorous field work, and its necessity was soon being publicized by figures such as Rivers and later, most significantly of all, by Bronislaw Malinowski. Over the decade ES investigated "multitudinous remains" of early Egyptians, aiming to classify them anatomically, "reconstructing their history and correlating it with that of mankind in other parts of the world. His initial conclusion from these Egyptian investigations was that, around 3000 BC, Egypt had been invaded by an alien race with distinctive somatological traits. Having reached this conclusion, ES then attempted to trace the movements of the race from archaeological evidence".[30]

Naturally enough, as a result of his Egyptian work, ES soon developed a special interest in mummification and published many papers on the subject during the 1900s. His protégé Warren Dawson has left an account of ES's first major encounter in this area. In 1898 the mummy of the pharaoh Tuthmosis IV had been discovered in the tomb of another pharaoh (Amenophis II). Early in 1903 the Egyptian director-general of antiquities, the eminent Egyptologist Gaston Maspero, ordered that the mummy of the king be examined in the Cairo Museum: "Maspero had a great love of ceremony, and he invited a number of distinguished personages to witness the unrolling", including Lord Cromer, the pro-consul, and Howard Carter. (Carter was then chief inspector of antiquities in Upper Egypt and had discovered the tomb of Tuthmosis IV while directing excavations for the American Theodore Davis in the Valley of the Kings. He is famous of course for opening up the tomb of Tutankhamen in 1922.) The body was duly unrolled. Various experts pronounced on the condition of the mummy and ES was asked to make a medical evaluation:

> ES found it impossible, under such spectacular conditions, to make more than a superficial examination of the body, and in order to furnish 'une evaluation exacte' he requested Maspero to allow him to make a detailed examination in private. This having been arranged, ES made a minute scrutiny of the body, upon which he published a detailed report. He found it necessary, in order to estimate the Pharaoh's age from the condition of the epiphyses, to have an X-ray examination. At the time there was only one radiographical apparatus in Cairo, and this was in a private nursing-home. He has since described to me very amusingly how he (and, I think, also Mr. Howard Carter) took the rigid Pharaoh in a cab to the nursing-home. It was the first mummy ever submitted to X-ray photography.[31]

In an extraordinary gesture of confidence, and what would today be regarded as one of gross political incorrectness, Maspero invited ES to examine the mummies of the greatest kings in ancient history, most notably those discovered at Deir el-Bahri (1881) and the tomb of Amenophis II (1898). Over two years of intense research ES asked for and was granted permission to examine numerous mummies from a number of dynasties, ranging in class from kings and priests to less exalted personages. At one stage Maspero transferred forty-four mummies from the museum to the medical school laboratory. This research enabled ES to decipher the intricate embalming techniques used, and to discern differing techniques used in differing periods. He continued his work intermittently during the following years and published a classic account

The Royal Mummies in 1912. By this time he was an acknowledged world authority on mummification.

ES combined his new interest in archaeology and ethnology with his primary interest in neurological research, continuing to publish important papers. Some purists have regretted that he was "sidetracked" into Egyptology. But, as his friend and mentor J. T. Wilson observed in an acute analysis, his Egyptian work actually enhanced his research in some directions:

> The anatomical school in Cairo offered an almost unlimited provision of recent brain material, and his studies of the Egyptian brain – more especially of the occipital region – not only revealed the 'Affenspalte' as an underlying feature of the human brain but led directly to his investigation, in 1907, of the topography of the central cortical areas and to the sharp differentiation of the visual area striata. His extensive study of the brain of both recent and ancient Egyptians was an important factor in the growth of his interest in physical anthropology and of an ever-widening outlook on the course of human evolution, which soon came to rank as one of the major interests of his life . . . His interpretation of the facts of the structure and evolution of the brain became clearer and riper as time went on. His 'Aris and Gale' Lectures of 1909 may be taken as a landmark of the progress of his neurological thinking, an illuminating interpretation of the essential factors in cerebral evolution and morphology. From this period onwards his attention was focused less on the intimate structural detail of the brain, but rather on the brain, human or mammalian, as an index of evolutionary status.[32]

Although he was under unrelenting pressure of work in Egypt (his letters are full of complaints that he didn't have time to do all the things he needed to do[33]), he had generous leave conditions (part of the "expat" system that operated.) Thus he was able to keep up his contacts with trips to Britain and elsewhere. He voyaged back to Sydney with wife and child in mid-1902 to visit family and university colleagues such as J. T. Wilson (with whom he continued a life-long correspondence). In May 1904 he attended an international commission of neurologists in London. He also attended a meeting of the British Association at Cambridge, and a year later he went to Geneva for the first international congress of anatomists. In 1906 he was in Dublin for the summer meeting of the Anatomical Society, and also visited colleagues in Glasgow and Edinburgh. In 1907 he attended meetings of the British Medical Association and the anatomical society held in England. He needed to cultivate people as he had his eye on a chair in the United Kingdom, or possibly America.

His last years in Egypt were preoccupied with his work as anatomical adviser to the celebrated Archeological Survey of Nubia. It was necessitated by the Egyptian government's decision in 1907 to raise the Aswan dam, which threatened to flood some of the world's most valuable antiquities. The archaeological rescue project under the control of Gaston Maspero, organized by Henry Lyons, used a team of experts including G. A. Reisner, and excavated over 20,000 burial sites. The results were published in a series of famous bulletins and reports in English and French. ES was responsible for the anatomy reports, a huge task. As Dawson records, when ES accepted the appointment he had no idea of the size of the task: "He envisaged a modest harvest for which a week's investigation each season would be adequate. But events turned out very differently, for the first season's campaign brought to light some six thousand bodies (skeletons and mummies). On his first arrival in Nubia in the second week of October [1907], Elliot Smith found over two thousand burials (the results of one month's excavating) awaiting his inspection. He soon found the task was beyond his unaided powers, and he made immediate arrangements for an assistant. He was most fortunate in securing the services of Dr. F[rederick] Wood Jones" (p. 45). Wood Jones, later professor of anatomy at Adelaide, Melbourne, and Manchester, became another long-term friend.[34]

In his reminiscences of Nubia, Wood Jones gave a rare insight into ES's personality:

> more than any man that I have ever met, he was indifferent to his surroundings . . . it might be said with truth that he carried his own environment with him . . . The only local incidents that affected him were the material objects, such as anatomical subjects, libraries and museums, that happened to be within his reach. It is certain that his Egyptian period changed and enlarged his outlook; but it was not, as many have supposed, because the romance of the land of the Pharaohs attracted him or had him under a spell. It was rather because Egypt furnished him with skulls and skeletons and mummies; and upon these things he was asked to report to Egyptologists.

This prompts the speculation that ES – like many scientists – may have had a degree of Asperger's syndrome (a form of autism). Wood Jones noted the stoicism and good humour with which ES endured the trying conditions in their weeks of camping out in the desert sands, flies and heat. But he also noted ES's rather mystifying lapses into apparent indolence when they were returning by the Nile to Cairo. Wood Jones deemed this a "splendid opportunity" to write up their notes: "For those three weeks he was everything that a good companion should be. He told amusing stories, discussed anatomical problems, detailed the shortcomings of contemporary scientific men: he did anything and everything that was amusing and pleasant – but he would *not* work. . . . Whilst he idled on the *Tawaf* upon the falling waters of the Nile, he was envisaging the wide ramifications into which his observations upon early cultural spread in the Nile Valley were inevitably leading him."[35]

ES's scientific reputation was now stellar, and this was recognized by his election to the Royal Society in 1907. Two years later the anatomy chair at Manchester University became vacant, and the university senate invited him to the post. He accepted and left Cairo in July. His Egyptian phase was over, but his obsession with Egypt was definitely not.

He determined to make a first-class anatomy school at Manchester, and during his ten-year stint largely succeeded.[36] The University of Manchester was at the time undertaking a lively expansion programme. The university included a galaxy of stars, including Ernest Rutherford (his old friend, the New Zealander who had won the Nobel Prize for chemistry in 1908, the man who established the structure of the atomic nucleus), the physicist Niels Bohr, the Australian-born philosopher Samuel Alexander, and the psychologist T. H. Pear who founded the psychology department there and was to become a leading propagator of Freud's psychoanalytic ideas. Pear has given a vivid account of the vibrant interchange of ideas that took place in "those happy days between 1910 and 1914":

> Occasionally we would gather – 'prayer meetings' they were called – in the evenings, when one member of a small, but never defined, circle would recite his scientific creed and answer questions about it. Rutherford, Niels Bohr, and Elliot Smith were the reciters who impressed me most. I expounded Freud's views, new and very exciting then.

In these discussions ES vigorously disputed Freud's belief in the "fundamental similarity of the human mind":

> So, Elliot Smith held, the view that when in widely separated parts of the world, similar beliefs, customs, habits, and techniques occurred, any explanation must rest on the fundamental similarity

> of the human mind, was vague and misleading. Did anyone, he asked, believe that the existence of Christian beliefs and ways of worship in distant parts of the world had little or nothing to do with migrating missionaries?

ES expounded his theories on "culture-bringers" and mummification at length. Against the critics who scoffed at ideas that primitive peoples could sail vast distances at one go, ES pointed out that such journeys could be undertaken piecemeal, with long journeys being undertaken in sections with long intervals in between voyages: "An Australian could hardly be expected to forget that strange men had come to New Zealand in big boats".[37] Rivers had contributed in two ways: (1) by showing that "primitive peoples" had experienced culture contact by means of sea voyages, and (2) by suggesting that even highly useful arts such as navigation that had been passed on to island inhabitants of the Pacific often "degenerated" or even disappeared in their later history.[38]

Later, during the First World War, Pear, ES and Rivers came together at the Maghull military hospital, near Liverpool, to treat mental patients. Pear and ES were to produce a book, *Shell Shock and its Lessons*, in 1917 and Rivers was to become a world authority on psychotherapy. Pear recollected how ES and Rivers would talk at nights, like priests, of mysteries:

> Gradually I pieced together the jigsaw of their references; 'the mother's brother', stone seats, megaliths, rags hung on trees, dragons, cranial deformation, circumcision, mummification and other heterogeneous items in the 'bag of tricks' [the culture clusters that ES alleged had traveled from Egypt]. They made world-maps showing the routes of the voyagers and the location of their alleged attractions: gold, pearls, and purple.[39]

W. J. Perry had been working for some time mapping global mining sites for precious metals and it appears that ES had absorbed this knowledge by around 1910. He was soon developing maps that correlated the locations of his Egyptian culture clusters (his "bag of tricks") with locations of precious metals and commodities.

During 1911 ES attempted to write a brief overview of his Egyptian researches to show how Egyptian history fitted into the broader evolutionary history of humankind (an abiding interest of his). This would bring together his specialist knowledge in the structure of the human brain and his amateur ethnology. The result was his popular book *The Ancient Egyptians* (1911). A major concern of the book was to suggest that "Armenoid" (or "Alpine" or "Slav") peoples had intruded into Egypt about 3000 BC. But, as he later recalled: " in the course of my investigations of these people, which their easily identified traits made possible, I became convinced that the rude stone monuments of the Mediterranean littoral and Western Europe were not really the most primitive stages in the evolution of architecture, but were crude copies of the more finished and earlier monuments of the Pyramid Age in Egypt, made in foreign countries by workmen who lacked the skill and the training of the makers of the Egyptian prototypes".[40]

W. J. Perry suggested in a later memoir that ES was drawn to the problem of megaliths, their structure, function and distribution, by a paper read to the Royal Anthropological Institute in 1910 by A. L. Lewis. It was about dolmens in France (megalithic tombs that featured large upright stones supporting a chamber roof). Lewis was unable to explain them: "His confession of failure to account for the facts cited evoked Elliot Smith's instantaneous reaction. Familiar with the structure, significance and history of the *mastaba*, the rock-cut tomb and the pyramid", he immediately suggested the solution outlined above, namely that the dolmens were crude copies of Egyptian pyramids.[41] ES began intensive research into the global distribution of megalithic monuments. He aimed to test the hypothesis that Egypt was the original inspiration for many of them (as well as for practices such as mummification). He claimed that dolmens found in parts of Europe had details of structure suggesting their derivation from Egyptian mastaba of the Pyramid era, a flat-topped structure at the top of the tomb.[42] Lewis, for one, was not impressed.[43]

By ES's own account, two events in May 1911 in Cambridge impelled him further along the heliolithic path. Whilst on his way to Cambridge to examine for the Natural Science Tripos he posted off to his publishers the manuscript of his *Ancient Egyptians*. When he reached Cambridge he called upon his friend Rivers to give him an account of what he had just done:

> he told me that my first incursion into ethnology was a flagrant defiance of all the current doctrines of that branch of study, and would draw down upon my head the most bitter opposition – a prediction that was amply fulfilled. However, he reassured me by telling me that he was actually engaged . . . on the task of writing his Presidential Address for the Anthropological Section of the British Association, in which he was making a full and frank recantation of his former acceptance of the orthodox ethnological doctrines. Although it was not until seven years later (1918) that Dr. Rivers went the whole way with me in recognizing the initiative of Egypt in the creation of civilization (*Psyche*, vol. III, 1922, p. 118), the fortunate circumstance of his change of opinion in 1911

> played a very material part in securing any hearing at all for my heresies.[44]

Rivers in his address made the striking point that, as a result of his expedition to Melanesia in 1908, he had been converted from the evolutionary paradigm then current to explanations of culture based upon racial mixture and the blending of beliefs and customs.[45] Malinowski later pinpointed this address as the start of the conflict in Britain between Diffusionism and Evolutionism.

That same day ES had another experience "of a very different nature that was destined to have far-reaching consequences, although at the time it was very disconcerting". In *Ancient Egyptians* (which he had just posted to his publishers) ES had clearly set out the physical traits of the aliens who (he believed) had made their way into Egypt about 3000 BC. This was the most important scientific finding in the book:

> When I entered the examination room in Cambridge, what was my surprise to see an example of this type [which had been chosen by his fellow examiner] to test the candidates' knowledge of racial peculiarities! Filled with curiosity, I consulted the Museum catalogue to discover from which of the geographical areas of the Ancient East enumerated in this book the specimen had come; but to my immense amazement I learned that it came from the Chatham Islands, near New Zealand, about as far distant from the Ancient East as was possible. For a time I was in some doubt whether or not I should recall the manuscript of this book, for the discovery of this skull seemed to destroy the very foundations of the argument set forth in it. However, further examination of the available craniological material revealed the widespread distribution of what in this book I have called 'Giza traits' not only in Polynesia, but also in the Malay Archipelago and at certain places on the southern Asiatic littoral.[46]

The evidence was beginning to suggest widespread movements of people analogous to those in the Mediterranean area. Even more startling, crania with similar traits were not uncommon on the Pacific coast of central and south America. He searched for cultural evidence to corroborate the physical evidence. This was provided by the distribution of megalithic monuments in India, Eastern Asia, Oceania and America. Hence, four months later, at the Portsmouth meeting of the British Association, ES put the cat among the pigeons, claiming that the sporadic distribution of megaliths west and east of Egypt, as far as Britain on one side and Japan and America on the other, was due to the influence – direct or indirect – of Egyptian civilization: "Small groups of people, moving mainly by sea, settled at certain places and there made rude imitations of the Egyptian monuments of the Pyramid Age".[47]

How had this arisen? At Portsmouth he explained that Proto-Egyptians in the pre-dynastic era had discovered copper[48] and had become skilled in making copper weapons, which enabled them to unite Egypt and to push into nearby Syria and Lebanon. They met and intermingled with the Armenoid people of northern Syria (who, having gained knowledge of copper weapons, then began conquering parts of neolithic Europe). In Egypt itself the people used copper tools, becoming expert carpenters and stonemasons. During the early dynasties "they ran riot in stone, creating the vastest monuments that the world has ever seen". Knowledge of these achievements spread to kindred peoples on the southern shores of the Mediterranean, southern Italy and Spain: "But it was the knowledge of the various kinds of monuments that the Egyptians were building, and not the skills nor the skilled workmen that spread". Thus it was that at about the time of the sixth dynasty a strong fashion of building stone monuments (dolmens, menhirs, cromlechs, rock-cut tombs, etc.) in the Egyptian-style began to spread to both west and east of Egypt.[49] This was the kernel of his *Ancient Egyptians*.

In this little book he used his massive expertise to describe the physical characteristics of the early Egyptians, or "Proto-Egyptians" as Arthur Evans had dubbed them. This people, he argued, contrary to some existing theories, had created early Egyptian civilization. It was an indigenous product, not the result of outside incursions. These people had "family likeness" to early Neolithic peoples in the whole Mediterranean littoral, the Iberian Peninsula, western France and the British Isles, north-east Africa, Persia and the Arab world. The genetic evidence suggested that they were "certainly the offspring of one mother", and probably – along the lines of Charles Darwin's theory – "out of Africa".[50] They were people of small stature, slender build, brown skinned with dark hair and eyes, in fact very like contemporary Egyptians and Arabs. ES's description of the "Mediterranean race" (a term popularized by the Italian anthropologist Giuseppe Sergi) sledge-hammered his readers with technical detail. Here is a sample:

> Over the whole of this wide domain the people were long-headed brunets of small stature, glabrous, and with scanty facial hair, except for a chin-tuft; with bodies of slender habit and a tendency to platycnemia, and *pilaster* and platymeria of the femur, and perforation of the coronoid fossa of the humerus. The skull is distinguished in all of these peoples by being long, narrow, ill-filled . . . , and a tendency to assume a pentgonoid (coffin-shaped) or ovoid form, when viewed

> from above; the eyebrow-ridges are poorly developed or absent; the forehead is narrow, vertical, smooth, and often slightly bulging; and the occiput is bulged out into a marked prominence of the back of the head. The forms of the orbits are either horizontally-placed ellipse or small circles . . . The cheeks are narrow . . . The nose is only moderately developed; it is small, and relatively broad and flattened at its bridge. The chin is pointed and the jaw very feebly built. The face as a whole is short and narrow; it is ovoid in form and straight . . . The whole skeleton is of slight and mild build, and has a suggestion of effeminacy about it. (pp. 58–59)

By contrast, the Armenoids, who intruded into Egypt at about the time of the Pyramid-builders and whose original home was probably in Asia, were long-bearded people, with solid build, larger and broader heads, square faces, longer and narrower noses, oblique orbits and different shaped jaws.

ES's book was founded on the work of scholars such as Reisner and Sergi, but he believed that he had advanced their hypotheses significantly. As he said: "The writings that embody the achievements of modern scholarship and fill the swollen shelves of our libraries will be searched in vain for any just appreciation of the influence exerted by Egypt's early culture on the nascent civilization of Europe and the world at large". Sergi had riveted attention upon Africa as the source of important elements in Europe's population and culture, "but the precise role played by Egypt, and the manner in which her influence was exerted, have never been explained" (p. 15). He believed he had solved some of the outstanding puzzles in this area, and the book was well received. But, as he also expected, and welcomed, his ideas would spark debate and encourage new research.

Ancient Egyptians focused upon North Africa, the Mediterranean and Europe, but his speculations also covered central Africa and western Asia. He began looking further afield, to India, South-East Asia, the Pacific and, ultimately, America. How had diffusion of culture actually worked? Rivers pushed ES further along the path of actual migration as the mechanism by which diffusion had occurred. ES had sent Rivers a copy of *Ancient Egyptians* hot off the press. Rivers had been working on physical movement of peoples for some time. He passed his views on to ES, who quickly took them up. As he wrote to Rivers on 14 October 1911:

> With reference to the migration question your remarks in June quite converted me to your point of view . . . Knowing nothing of the evidence on your side when I wrote the book I naturally took the more cautious line of assuming that the culture may have spread

> amongst kindred peoples without any great racial movements: but as the incoming of 'Armenoid' traits into Sicily, Italy and North Africa generally, is as patent as it is in Egypt, there is positive evidence for the movement of people: and in the light of your work, I have no hesitation whatever in speaking now of this movement as the explanation of the spread of culture – as in fact I did in my hurried remarks at Portsmouth.

ES then took up his cause "with the ardour of a missionary", preaching his gospel at learned societies and at numerous public meetings in places like Manchester, Glasgow, Edinburgh and Belfast: "He provoked the attacks of orthodox anthropology at the meetings of the British Association in 1912, 1913, 1914 and 1915, and he always looked back with intense delight and amusement upon the (sometimes very heated) discussions that arose out of his remarks".[51] There is no doubt that ES could be provocative. Graeme Pretty suggested (in 1969) that ES's "diffusion heresy" raised hackles less because it postulated cultural diffusion (which had long been partially accepted by "evolutionist" anthropologists) than because of its insistence on transmission through migrating bands, with racial affiliations to ancient Egyptians. Pretty added: "It must not be forgotten however that much of the opposition to ES was more of a personal reaction to his pugnacity and sharp treatment of his intellectual opponents, and the memory of it survives".[52] Solly Zuckerman later commented: "Controversy and criticism are necessary parts of the scientific process, and . . . my own view is that much which has been published as contributions to our knowledge of human ancestry, but which later proved worthless, would have enjoyed a shorter life had there been more Elliot Smiths in the world". He added that "it would be wrong to say – as many have said – that he shut his eyes to new evidence or that his mind was closed to new arguments".[53]

Dawson (who was to carry on ES's work on mummification in the 1920s) recalled that ES went out of his way to circulate, not only his own papers, but those of his harshest critics. He "always generously admitted that many of his opponents' statements against his heresy had done more to establish his point of view than anything he himself could say or write of it".[54] In other words he enjoyed keeping the controversy alive as it gave oxygen to his theory. It would prevail (he believed) because it was supported by convincing scientific evidence, industriously and painstakingly collected from an impressive range of sources.

CHAPTER

2

Migrations of Early Culture (1915)

In February 1915, just as the Great War was getting under its gruesome way, ES read a major paper to the Manchester Literary and Philosophical Society on mummification and the spread of customs. It was his first major statement of his heliolithic theory. Very rapidly he had it published in Manchester, London and New York as a short book.[1] As he said, apologetically, it had been "crudely flung together" (p. v). It was in fact a shot in the "anthropology wars" that soon broke out over diffusionism. As we shall see, it was fired after ES had been involved in a skirmish at the British Association meeting that had been held the year before in Australia. ES was provoked to demolish his critics in Melbourne.

The book was entitled *The Migrations of Early Culture*. ES's aim was to show that a body of evidence collected by authorities in the field could be more convincingly explained by his diffusion of culture theory than by the current ideas of anthropologists, "an appeal to ethnologists to recognize the errors of their ways and repent" (p. v). ES was already in active revolt against prevailing ideas of "the psychic unity" of humankind and the independent invention of customs and practices, as epitomized by the legacy of the German ethnologist Adolph Bastian.[2] ES was aware of Continental revisionist ideas that rejected Bastian, but he was sceptical that some of them went too far in the other direction – most notably the "Viennese School" associated with the Catholic priests Fritz Graebner and Wilhelm Schmidt (a philologist) and the Swiss ethnologist Montandon (p. 18).

Their problem (he thought) was that they placed too much emphasis upon material culture (artifacts such as bows and spears). Following Rivers, he argued that more reliable conclusions could be drawn using broader concepts of culture, looking for instance at social organization "as supplying the most stable and trustworthy data for the analysis of a culture-complex and an index of racial admixture" (p. 19). ES's emphasis

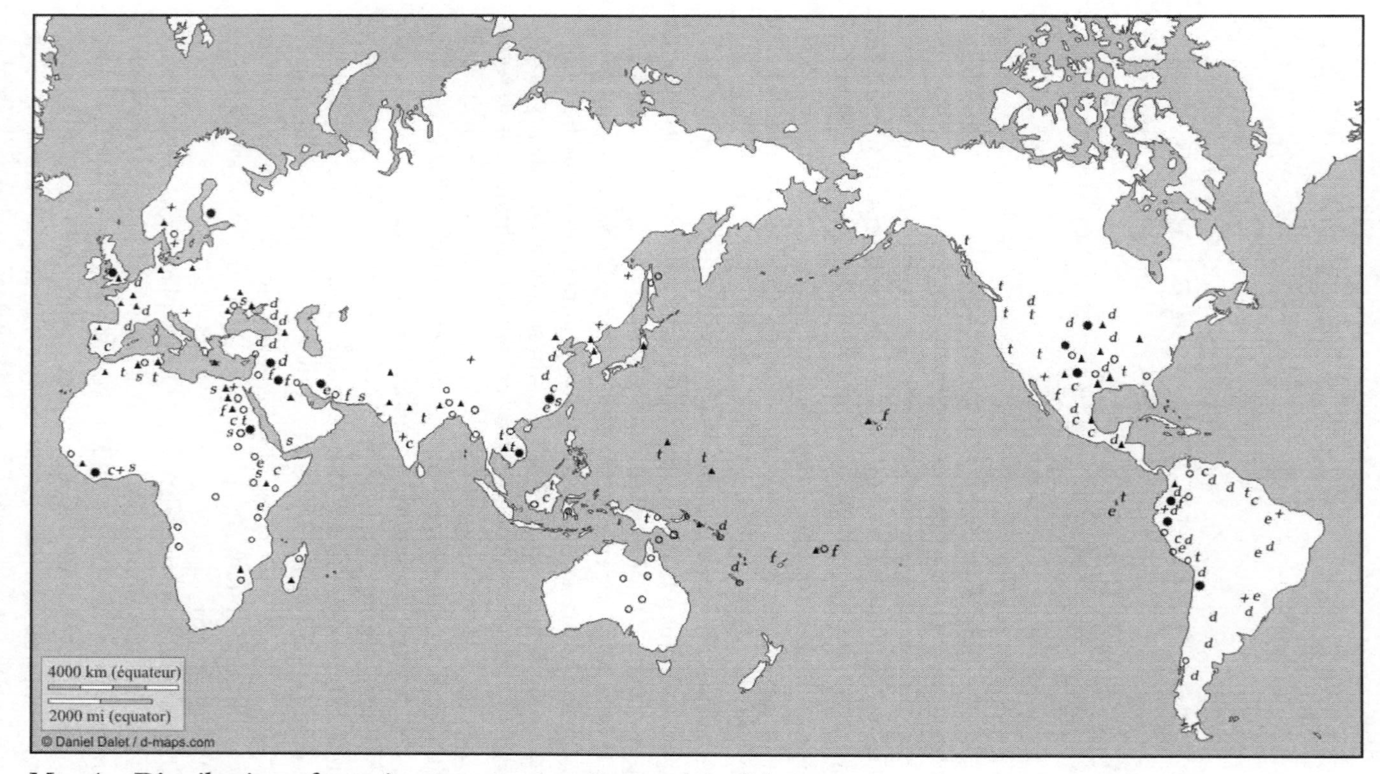

Map 1 Distribution of certain customs, practices and traditions
From page 2, *The Migration of Early Culture* (1915) (Redrawn by Claire Reeler and Nabiel Al Shaikh, 2011)

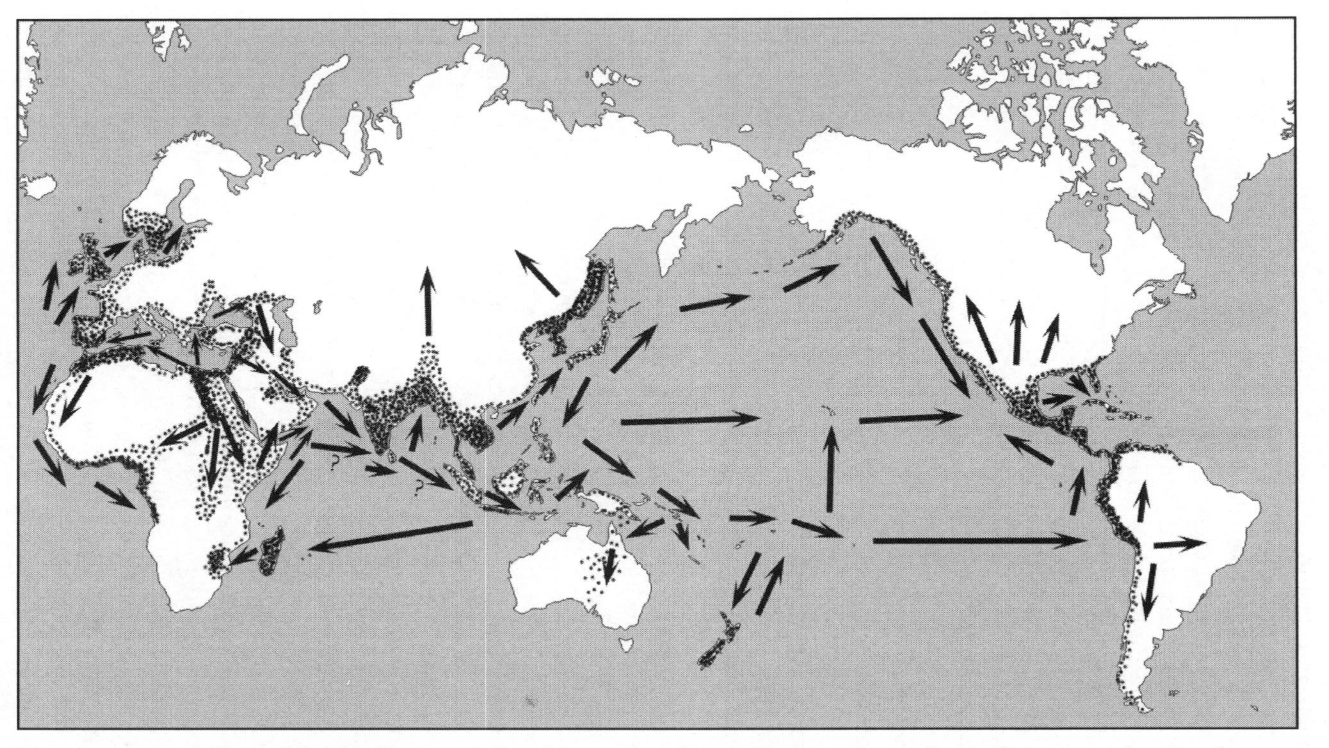

Map 2 Areas affected by "heliocentric" culture-complex, indicating hypothetical routes taken in the migration of culture-bearers
From page 14 of *The Migration of Early Culture* (1915) (Redrawn by Claire Reeler and Nabiel Al Shaikh, 2011)

upon "culture-complex" and holistic factors tends to negate criticism that the diffusionists focused almost exclusively on particular cultural features while neglecting factors such as functionality or social structure. He aimed to look at a cluster of "peculiar practices" and customs that related to peoples' deep rooted beliefs. This approach was by no means new. The idea of diffusion of clusters or complexes of cultural traits had been advanced in the nineteenth century, most notably by the Kulturkreise group led by the German ethnologist Friedrich Ratzel (1844–1901). Ratzel had suggested that migration of peoples was a key factor in spreading cultures from some common source or epicentre, "an elaborate theory that had cultures everywhere developing as a result of overlapping bundles or complexes of traits carried from some heartland in great waves or circles".[3]

From his own background in Egyptian studies and using global anthropological and archaeological evidence, ES chose a cluster of practices that he believed stemmed from Egyptian origins and were to shape the history of civilisation. From Professor Brockwell he borrowed the term "heliolithic culture" to describe a culture that included sun-worship and megalithic building. Customs that were fortuitously, but intimately, interlinked formed a definite culture-complex nearly 3,000 years ago. They spread along the coastlines of a great part of the world "stirring into new and distinctive activity the sluggish uncultured peoples" who were subjected to this "exotic leaven" (p. 1).

We may note the anti-populist assumptions here: masses of people tend to be culturally inert and traditionalist unless "leavened" by elitist innovations. It is possible to interpret such assumptions as the product of the age, one of a decline in individualism, loss of faith in progress and the rise of authoritarian doctrines. However, as a democratic Australian, ES was not temperamentally inclined to authoritarianism. Nor did he succumb to the biological determinism that flourished at the time and that underlay much racist ideology in the west. As a biological scientist ES was convinced that *Homo sapiens* were one species, and no one subspecies or "race" was genetically superior to another. (As we shall see he became a fierce opponent of the Nazis' Aryan race theory during the 1930s.) However in terms of cultural evolution, he followed Darwin's line that some groups had greater social efficiency than others (although this was always a flexible situation, with "top dog" groups being constantly supplanted by up-and-coming groups). Some cultures, most notably the early Egyptians, had taken advantage of uniquely fortuitous environmental and historical circumstances to build pioneering civilizations. Creative cultures spread almost infectiously to other, less adventurous, cultures. While he granted that all cultures had the potential to expand and diversify, ES undeniably inclined to a conservative assessment of human creativity, but it was not a view that rested upon genetic or racist assumptions. This needs to be emphasized as his diffusionism has sometimes been dismissed as racist.

ES was much influenced by scholars such as Augustus Lane-Fox Pitt Rivers (1827–1900), who did ground-breaking work in the late 1860s and early 1870s on megaliths and early navigation; Miss A. W. Buckland, with her work on serpent worship and tattooing in the 1870s; and Zelia Nuttall, with her studies in the 1900s on Mexico and the use of shell-fish for dyeing. Another influence was W. H. R. Rivers, on whom more anon. From his own study of mummification ES found that its geographical distribution exactly corresponded to that of a group of other "peculiar practices". They included the building of megalithic monuments, sun worship, use of the swastika symbol, serpent worship, cranial deformation, ear piercing, the creation story of the deluge, and tattooing. ES produced a spectacular global map by plotting the distribution of these customs using symbols (Map 1). It showed a thick clustering around Egyptian North Africa and the Middle East (in his theory the starting point of a mass cultural diffusion). The symbols are then pictured in connected lines and clusters spreading in what look like cultural trade routes that encompass large parts of southern Europe up to Britain, down into Africa, across to India and South East Asia, to New Guinea, Australia and Pacific Islands, up to China and Japan, and finally across to the Americas, with heavy concentrations in Central America and the west coastal regions of South America (Map 2).

It was an impressive piece of work. Although ES encountered heavy weather convincing cynics, he believed that his findings were totally incompatible with the doctrine of independent invention. How was it possible that such intricately interconnected practices, showing remarkable identities of structure even when it came to trivial and inessential features, could have been independently invented? He had no faith whatsoever in explanations based upon stage theories of evolution or vague assumptions about common human talents or instincts.[4]

Stage, or unilinear, theories of evolution had dominated ethnological thinking in the second half of the nineteenth century. (In the usage of the day ethnology included archaeology and anthropology.) Such theories asserted that all human cultures necessarily evolved through a number of stages or phases, evolving from simple to complex, from ape-man hunter-gathering to modern industrial nations. (Herbert Spencer's abstract theorizing along these lines had enormous impact.) ES objected to this intellectual straitjacket, on the grounds that it did not reflect cultural realities and – of major concern – that it did not allow much room for diffusion of cultures. ES was an evolutionist but he didn't think evolutionary principles applied directly to cultural evolution and certainly not in this deterministic way. Unilinear theory was under challenge even at this time and increasingly so during the twentieth century. It was finally abandoned in favour of more relativistic approaches that emphasized the need to examine cultures and peoples in their own context and terms. ES had no problem with this, as long as the focus was not so narrow that it discouraged the possibility of external influences. He felt that some of his opponents had simply swung from one extreme to the other.

ES was struck by work done in the 1900s by Zelia Nuttall showing remarkable parallels between Eastern Mediterranean culture in Phoenician times and that of pre-Columbian America. She noted closely associated features in both Old and New Worlds: the purple industry and weaving skills; the use of pearls and conch-shell trumpets; copper, silver and gold working and trading; the tetrarchial form of government; the conception of "Four Elements"; and the cyclical form of calendar. Nuttall thought it at least possible that such Egyptian-style customs had been transmitted by bands of Mediterranean sea-farers. This possibility (she thought) was bolstered by indigenous American traditions that strangers of superior culture had brought such knowledge from distant parts. She tended to be dismissed as a woman with dotty ideas.

By the early twentieth century knowledge was rapidly expanding about prehistoric migrations of peoples (for instance in the Pacific) and the surprising scale of sea travel. W. H. R. Rivers emphasized the migrations of culture-bearers who exerted profound influence upon "uncultured" populations. He observed how (for example) the "fantastic" steering arrangement used by the Egyptians 3,000 years ago survived into modern times in the Hoogly and other Indian rivers.

What motivated the wanderings of the Egyptian travelers? During his exploration of megalithic trails from Egypt into the Pacific, ES had come across the work of William James Perry (a former student of Rivers), which suggested that the search for precious commodities may have been the answer. Rivers later recollected how Perry had shed light on the issue at a crucial time:

> ES had, at the beginning of 1915, mapped out the areas which he knew, or inferred, to have been the seats of megalithic influence. He sent Perry a copy of this map. About the same time he informed him that Mrs Zelia Nuttall had claimed that the founders of American civilisation had shown a special appreciation of pearls and precious metals. A chance examination of an economic atlas enabled Perry to combine these two items of information, for he saw that the distribution of pearl-shell in the Pacific Ocean agreed so nearly with ES's megalithic areas that the presence of pearls would provide a sufficient motive for the settlements in these

regions. Further examination showed that, in inland areas, the chief motive for the megalithic settlements was supplied by the presence of gold and other forms of wealth. He therefore put forward the view that the motives which had led the ancient so far afield was precisely that which acts as the essential stimulus to our own migrations, viz., the search for objects needed to satisfy human needs, material, aesthetic and religious.[5]

Having shown the spread of peoples and culture from the Mediterranean littoral to western Europe, ES proposed next to show a spread from the southern Asiatic coast into the Pacific. He had planned to advance his theory mainly by work on megalithic monuments but supplemented by reference to mummification. This strategy was to be changed by events associated with the British Association meeting in Melbourne in August 1914.

Before the meeting (in July) he examined a Torres Strait mummy held in the Macleay Museum at his old medical school at the University of Sydney and studied the literature on embalming methods used in the Papua area. ES was of course a world authority on the technical details of Egyptian mummification. He was startled to discover that the Papuans had been using remarkably similar techniques: for instance, the incision for eviscerating the body was made in the flank or perineum (the flank incision in precisely the same situation that was distinctive of the XXIst and XXIInd Dynasty methods in Egypt); the viscera was thrown into the sea; the body was painted with a mixture containing red ochre, the scalp painted black and artificial eyes were inserted; "But most remarkable of all, the curiously inexplicable Egyptian procedure for removing the brain, which in Egypt was not attempted until the XVIIIth Dynasty – *i.e.*, until its embalmers had had seventeen centuries experience of their remarkable craft – was also followed by the savages of the Torres Straits!" (p. 22).[6]

The zoologist D. M. S. Watson was present at the Macleay Museum. He left a vivid account of events, showing that ES was clearly not expecting to find Egyptian-style methods of mummification in Papua (but rather local methods of smoke-drying and soaking in sodium carbonate):

> The mummy we examined was the body of an adult man, shrunken but well preserved, lashed down at full length to a ladder-like framework of poles. As soon as we took it out of the museum case, Elliot Smith pointed out that all the finger- and toe-nails were carefully tied on with a binding of thread, and with obvious excitement told us that this was a custom found in Egyptian mummies of a defi-

nite date, and that this had the definite function of keeping the nails attached during the long soaking. [ES then closely examined the surgical operations that had been performed upon the body]. The whole examination was most dramatic: the immediate discovery on the Torres Straits mummy of custom after custom of the Egyptian mummifiers, as they were described to us by Elliot Smith, made a vivid impression on my mind . . . [Later in the day] Elliot Smith developed to me his interpretation of the facts we had seen. He pointed out that the preservation of a human body was so long, difficult and gruesome a process that no people would undertake it unless they were driven by precise and definite religious beliefs which made it necessary to preserve a man's body in the form it had in life in order that his soul should have a habitation. Furthermore, the idea that it was possible to preserve a body permanently could only have arisen from the observation of the accidental preservation of bodies by dessication through burial in hot dry sand; and that as a matter of historical record the earliest known mummies are Egyptian; that in Egypt predynastic men, buried in shallow graves in the sand, without coffins, were in fact well preserved; and that the basis of Egyptian religion lies in the power of the soul to reinhabit its body and to survive only as long as it can do so. As none of the conditions which led to the discovery of mummification by the Egyptians exists in the Torres Straits, it is evident that the art of mummification could not have arisen there but must have been introduced either by a migration of the actual people who inhabit that region or by some spread of culture.[7]

ES (on 30 August 1914) also examined some mummies in the Brisbane museum that reinforced his Sydney observations. What impressed him especially was that the Papuans had apparently reproduced complex techniques used by the Egyptians (presumably connected with their belief systems) when much simpler methods would have served their purpose. Aside from the method of mummification there were also stunning similarities in the special treatment of the head, the use of masks, the making of stone idols and other funerary practices. These had been observed by J. L. Myres and A. C. Haddon during their famous Cambridge expedition to the Torres Straits in 1898. The evidence also led him to conclude that the cultural wave that (in his opinion) brought such expert knowledge to the Torres Straits could not have started before the ninth century BC (as such procedures were not invented in Egypt until the New Empire). Graeme Pretty's re-examination of the Macleay mummy in the 1960s basically confirmed ES's surgical analysis.[8]

When ES called attention to these facts at the Melbourne meeting[9] he was astonished, and angered, to find them treated with disdain by leading anthropologists. Ironically Myres and Haddon led the charge, despite their work in the Torres Straits. Myres, who was to become a power broker in the Royal Anthropological Institute, chaired the discussion. He asked what was more natural than that people should want to preserve their dead, remove more putrescible parts, cut into the flank, or paint their dead with red-ochre. Elliot Smith was appalled:

> The claim that it is quite a natural thing on the death of a near relative for the survivors instinctively to remove his viscera, dry the corpse over a fire, scrape off his epidermis, remove his brain through a hole in the back of his neck, and then paint the corpse red is a sample of casuistry not unworthy of a medieval theologian. Yet this is the gratuitous claim made at a scientific meeting! (p. 25)

He attacked Myres's ignorance of anatomy. This was probably the beginning of the frosty relations that marked relations between the two men for more than two decades, and was often played out in the Royal Anthropological Institute. Haddon objected that there were no links between Egypt and Papua to indicate the spread of the custom of mummification. ES wrote his *Migrations of Early Culture* to map out such routes, and to demonstrate the linkages between mummification and other practices, something his critics had entirely ignored in Melbourne.

ES found further links in the "cultural chain" in the research of Rivers and Perry. These three were to become the axis of the diffusionist school (much maligned by contemporary and later ethnologists). *Migrations of Early Culture* cited a number of "epoch-making" research papers by Rivers. Among other things (including how useful arts could be lost in the spread of culture) Rivers had "analysed and defined the characteristic features of several streams of culture which flowed from Indonesia into the Pacific" (p. 31). He published these ideas in his influential *History of Melanesian Society* (1914). But ES was even more excited by Perry's patient research on Indonesia: "With remarkable perspicuity he had unravelled the apparently hopeless tangle into which social organization of this ethnological cockpit had been involved by the mixture of peoples and the conflict of diverse beliefs and customs" (p. 30). His finding was that there had been immigration into Indonesia (from India and the West) of a people who introduced megalithic ideas, sun-worship, phallism and other practices which to ES were "heliolithic", including the practice of mummification.[10]

He and Perry spent months in detailed discussion of these issues, and the two were to become close. Almost certainly ES pushed the younger man further along the heliolithic path. Perry was as yet largely unpublished. Originally from the City of London School, he had come to Selwyn College, Cambridge, to read mathematics. His interest in anthropology had been sparked when he attended lectures by Rivers and A. C. Haddon. He began researching Indonesia in his spare time from school mastering. At Rivers's suggestion, he learnt Dutch in order to read sources. After his book *The Megalithic Culture* appeared in 1919, he was offered a readership in comparative religion at Manchester. This was no doubt engineered by ES and involved teaching courses in ethnology in the psychology department (like Rivers he developed a strong interest in psychology). In 1923 he joined ES as reader in cultural anthropology in the Department of Human Studies that had been set up by ES at London University. Perry was to make his name with works such as *The Children of the Sun* (1923), *The Growth of Civilization* (1924), and *The Primordial Ocean* (1935). An amiable personality, Perry could be formidable in debate.[11]

Meanwhile back in 1915, *Migrations of Early Culture* concluded with a grand sweep. A great migration of the ancient "heliolithic" culture-complex began about 800 BC Its reach extended not only to the Mediterranean and Europe but to China and Japan. Passing eastwards the culture-complex reached the Persian Gulf, strongly "tainted" with the influence of north Syria and Asia Minor; "and when it reached the west coast of India and Ceylon, possibly as early as the end of the eighth century BC, it had been profoundly influenced not only by these Mediterranean, Anatolian and especially Babylonian accretions, but even more profoundly with East African modifications" (p. 133). From Indonesia it was carried far out into the Pacific and eventually reached the American coast. Here it bore fruit in the development of the great civilizations on the Pacific littoral and isthmus.

ES and Perry defended their thesis at the Manchester meeting of the British Association in 1915. They argued that many distinctive Egyptian practices, possibly carried in part by the Phoenicians, had been taken to the coastlines of Africa, Europe and Asia, and later to Oceania and America. Perry elaborated his theory of the pulling power of precious metals, looking at the association of megaliths and mine workings or gold bearing regions and pearl fisheries, noting the similarities of techniques for smelting or refining metals. Rivers supported his friends and explained that he had modified his original opposition to their position after ES's find of the Torres Straits mummy.[12]

The pair was subjected to "severe criticism" (according to *Nature*'s report of the meeting). Arthur Evans attacked their methodology. The eminent Egyptologist Flinders Petrie required greater precision in dating the facts presented. (Dating was always a problem for the diffusionists and it was not until carbon dating that chronologies could be more accurately presented. Petrie and ES had already clashed over the interpretation of Egyptian mummies, and Petrie noted in his diary "much disgusted" after listening to ES and Perry.)[13] Charles G. Seligman (a close friend of Rivers) had already in his opening address cogently pointed to methodological problems involved in tracing ideas and customs to Egyptian (or any) source. He pointed out that general resemblances (for example, in forms of social organization and belief, such as matrilineal descent or cult of the dead) could not be admitted as good evidence: "I do not think that in the present state of anthropological science even extreme or unusual beliefs and devices (which at first sight seem so strikingly convincing) should be considered as proof of common influence".[14] The battle lines were being drawn.

CHAPTER

3

Dragons and Critics (1915–1920)

ES began to intensify his reading into key elements of his thesis, despite the distractions of the world war and the heavy demands of his teaching and research in anatomy at Manchester University. Convinced that the Phoenicians were key carriers of Egyptian culture, he read French and German archaeologists on the Phoenicians and immersed himself in early travel literature. He found a rich cache of material on maritime expeditions in Wilfred Schoff's translation of the first-century Greek work *The Periplus of the Erythrean Sea*, which dealt with travel and trade in the Indian Ocean. He read translations of Chinese works on the Chinese and Arab trade, and Terrien de Lacouperie's *Western Origin of the Early Chinese Civilization* (1894), as well as material on India and the Pacific Islands. Zelia Nuttall's comparative study of new and old worlds, and especially her work on Mexican religious and astronomical knowledge, continued to inspire ES.[1] Will Perry was also working tirelessly on diffusionist themes, and the two men traded information and ideas in repeated conversations at this time. A few years later ES would describe Perry's researches as "the most potent factor of all in shaping my views". These researches, he said, were "converting ethnology into a real science and shedding a brilliant light upon the early history of civilization".[2]

In March 1915 ES gave a lantern slide lecture on Ancient Egypt at the John Rylands Library in Manchester. His heliolithic ideas were beginning to create interest, not least in the United States, where they riled American anthropologists. They were most resistant to the concept of an old world influence on native American cultures, something that ES put down to nationalism. His critics put it down to their superior factual knowledge. ES aimed his lecture at a more popular audience, appealing to their "common sense" in support of the proposition that "the fundamental constituents of all civilizations spread from one centre". He projected lantern-slide images of tombs, temples, mummies and his global map plotting the spread of Egyptian practices. Over the next year he worked up this material into a more scholarly paper that took further the ideas of his *Migrations of Early Culture*.[3]

By this time ES was arguing the line that it was in fact easier to study the easterly spread of Egyptian influence from the Mediterranean – where conflicts between rival cultures made it difficult to decipher the Egyptian story "in its much scored palimpsest" – to areas "where among less cultured peoples it blazed its track and left a record less disturbed by subsequent developments than in the West" (p. 48). He was already being battered by classical archaeologists and scholars who disputed his readings of Mediterranean history. But his Eurocentric approach and assumptions of western cultural superiority, although the norm at the time, risked criticism by scholars of Indian and east Asian history, not to mention parts east of Asia.

According to the heliolithic narrative, key features of Egyptian culture spread from the west to help form the basis of the civilizations of India and the Far East. Currents spread also to the Malay Archipelago, Oceania and America. The carriers of culture were mariners who from around 800 BC began trading between the eastern Mediterranean and India. This went on for many centuries. The source of the "highly complex and artificial" culture that spread was, of course, Egypt, although ES was willing to concede a southern Mesopotamian connection via the Sumerians (who had strong supporters among archaeologists). The Phoenicians were the obvious candidates as carriers of culture, given their legendary status as seafaring traders from ports such as Sidon and Tyre (well-known to people through the Old Testament).

The picture was by no means a simple-minded one of direct Egyptian influence. Egyptian culture itself was not monolithic. As it was exported by various peoples and groups it was coloured by their respective values and absorbed elements of the local cultures it encountered. There were important accretions and modifications from the Phoenician world of the central-east Mediterranean, and also from East Africa, the Sudan, the Arab world and Babylonia. If Egypt was the "leaven" that helped to start early Indian civilization, then it was also true that a diverse Indian culture in turn "leavened" the eastern littoral of Asia and Oceania. Finally "the stream, with many additions from Indonesia, Melanesia, and Polynesia, as well as from China and Japan, continued for many centuries to play upon the Pacific littoral of America, where it was responsible for planting the germs of the remarkable pre-columbian civilization" (p. 49).

What were the signs of Egyptian influence? We have already seen that ES identified a remarkable cargo of "extraordinary practices and fantastic beliefs" that spread abroad and showed up impressively on his global map. Key among them was the practice of mummification "characterized by a variety of methods, but in every place with remarkable identities of technique and associated ritual, including the use of incense and libations, a funerary bier and boat, and certain peculiar views regarding the treatment of the head, the practice of remodelling the features and the use of statues, the possibility of bringing the dead to life, and the wanderings of the dead and its adventures in the underworld". Also there was the building of megalithic monuments conforming to recognizable Egyptian types and associated with identical traditions and beliefs; sun and serpent worship using symbols representing "an incongruous grouping of a serpent in conjunction with the sun's disc equipped with a hawk's wings"; and a string of other practices (pp. 49–50), some of which are discussed below.

ES's chronology of Egyptian influence was roughly this. By about 2800 BC, or even earlier, there was significant maritime intercourse between Egypt and peoples scattered along the eastern Mediterranean coasts, extending to the Minoans in Crete and even to the ancient civilization of the Sumerians at the head of the Persian Gulf. Egyptian megalithic culture had its impact upon the Minoans (witness their use of stone building methods and the making of rock-cut tombs). But as it became an advanced culture in its own right, Minoan civilization resisted "alien" influences. It was not until the Minoan state had fallen and Egyptian power had begun to decline that the free-minded Phoenicians stepped in. They had the knowledge of the Egyptians, the Minoans and the peoples of Levant and they had exceptional maritime skills. Thus, from about 800 BC for three to four centuries, they became the great intermediaries between the nations of antiquity. In the course of trade the Phoenicians "did not scruple to adopt [those nations'] arts and crafts, their burial customs, and even their gods. In this way was inaugurated the first era of really great sea-voyages in the world's history" (p. 53). In many ways this era surpassed even the great age of European exploration of the fifteenth and sixteenth centuries, given the vast global influence of the Phoenicians. They pioneered the use of gold as a trading currency, and the use of astronomy for navigation. (Zelia Nuttall had documented their use of the pole star for navigation.[4])

Their appetite for gold was legendary, but so also their search for precious stones and pearls. As Perry had noted at the Manchester meeting of the British Association in 1915, so thoroughly did they, and their pupils and imitators, accomplish their pearling mission that the Western Australian site of Broome was the one pearl-field to escape their exploitation. Perry had been assiduously studying the localized areas where distinctive "Egyptian" features occurred, especially megalithic structures, terraced irrigation, sun-worship and mummification. These were precisely the places (he claimed) where there were ancient pearl fisheries and mine-workings that used techniques identical to those employed by the Egyptians (the use of a blast over the mouth of the furnace, metal roughly smelted in open furnaces and then refined in crucibles, etc.). He speculated that the operators were probably Phoenicians who established settlements to exploit available sources of wealth.

According to ES, the Phoenicians brought heliolithic culture to pre-Dravidian India: "temples of New Empire type, dolmens which represent the Old Empire type, rounded tumuli which might be regarded as Mycenean, and seven-stepped stone Pyramids as Chaldean, modifications of Egyptians pyramids; and if the monuments farther east are taken into consideration, the blended influences of Egypt, Babylonia, and India become even more definitely manifested" (p. 66). Mummification and ship-building designs were also taken from Egypt. China also was vastly influenced. ES even suggested that Chinese writing derived from the Egyptian hieroglyphic system of writing.[5] This ran counter to the prevailing view of Sinologists, who believed that Chinese writing had much more ancient origins.

From India a series of migrations carried megalithic customs and beliefs east to Burma, Indonesia, China, Japan, the Pacific, and eventually to America "where there grew up a highly organized but exotic civilization compounded of the elements of the Old World's ancient culture, the most outstanding and distinctive ingredients of which came originally from Ancient Egypt" (p. 69). There was already convincing evidence that "Asiatic civilization" had reached America "partly by way of Polynesia, as well as directly from Japan, and also by the Aleutian route" (p. 70). The seafaring feats of the Polynesians and the sturdiness of their craft had convinced the sceptics that long voyages and migrations were feasible.

ES listed the elements of megalithic culture that had reached American shores in pre-Columbian times: practices of mummification (modified by oriental and Polynesian influences), rituals using incense and libations, beliefs concerning the soul's wanderings in the underworld, corroborative details showing Ethiopian, Babylonian, Indian and other influences, the use of idols and stone seats, the belief that humans and animals could dwell in stones, the story of the deluge, the divine origin of kings (children of the sun or sky), and the incestuous origin of a "chosen people". ES's list seemed endless:

> the worship of the sun and serpent-worship. Circumcision, tattooing, piercing and distending the earlobes, artificial deformation of the head, trephining [use of a surgeon's saw to operate upon a skull or organs], weaving linen, the use of Tyrian purple, conch-shell trumpets, a special appreciation of pearls, precious stones, and metals . . . terraced irrigation, the use of the swastika symbol, beliefs

> regarding thunder-bolts and thunder teeth, certain phallic practices, the boomerang, the beliefs regarding "heavenly twins", the practice of couvades,[6] the custom of building special "men's houses" and the institution of secret societies, the art of writing, certain astronomical ideas, and entirely arbitrary notions concerning a calendrical system, the subdivisions of time, and the constitution of the state – all these and many other features of pre-Columbian civilization are each and all distinctive tokens of influence of the culture of the Old World upon that of the New. (p. 71)

In August 1916 ES published the kernel of his argument in a short paper for the prestigious journal *Science*.[7] In it he reiterated his belief that no theory of independent invention could possibly explain his cluster of "strange ingredients . . . compounded in a definite and highly complex manner to form an artificial cultural structure". This was because chance had played such a large part in building up this structure in its original home in Egypt. A theory of evolution of customs might possibly explain the origin of some items, but it was totally inconceivable "that the fortuitous combination of hundreds of utterly incongruous and fantastic elements could possibly have happened twice" (p. 194).

ES was abrasively dismissive of countervailing opinions. He gave a thumbnail sketch of anthropological historiography that was undeniably simplistic. It went like this. Some forty to fifty years earlier anthropology had taken the common sense view that diffusion was normal, that there had been a spread of customs from the old to the new world. However a revisionist trend set in. It was based on the "rigid dogma" that, because of the inherent similarity of human minds across the globe, "similar needs and like circumstances will lead various isolated groups of men in a similar phase of culture independently one of the other to invent similar arts and crafts, and to evolve identical beliefs". The modern generation of ethnologists "has thoughtlessly seized hold of this creed and used it as a soporific drug against the need for mental exertion. For when any cultural resemblance is discovered there is no incentive on the part of those whose faculties have been so lulled to sleep to seek for an explanation: all that is necessary is to murmur the incantation and bow the knee to a fetish certainly no less puerile and unsatisfying than that of an African negro" (pp. 190–191). By this mode of thinking a modern day witness to the global spread of the steam engine would be obliged to put this down to a series of independent inventions. History and psychology were fatal to the idea of independent invention. Originality was the rarest of human faculties; while biological psychology showed that human instincts were extremely generalized, and not the highly specialized instincts that underlay the theories of modern ethnologists (ES was bafflingly vague on the supposed links between independent invention theory and concepts of special instincts). In a general insult ES accused ethnologists of refusing to accept the obvious meaning of facts, and instead resorting to "childish subterfuges" (p. 191). This was not calculated to win friends and influence people, and refutations soon began to rain in.

The American ethnologist Alexander Goldenweiser of Columbia University challenged ES to name one ethnologist who attributed similarities in cultures to the working of special instincts.[8] As to "psychic unity", Goldenweiser declared it "but a substratum, a universal common denominator" without which no culture could exist. It was manifested no less in mechanisms of cultural diffusion than in those independently invented:

> In neither case does 'psychic unity' become an explanatory factor. If there is such a thing as explanation in history, then the complete reconstruction of the historic event is the explanation the ethnologists would demand, in the case of diffusion as well as in that of independent development. (p. 532)

In a point that was to be frequently made, Goldenweiser argued that diffusion enjoyed a methodological advantage "for whereas diffusion can be demonstrated, independent development does not, in the nature of the case, permit of rigorous proof". If you argued for independent invention you were negating diffusion, "a negative based on negative evidence, absence of proof of diffusion". Diffusionists could always claim that at some time diffusion had occurred. Such a claim was unanswerable. Goldenweiser argued that it was preferable in contested areas for independent invention to be assumed until diffusion was proven "or, at least, made overwhelmingly probable" (p. 533). His own reading of history was that it proved beyond a shadow of a doubt "that independent inventions do occur as well as that originality, while rare in its most pronounced forms, is in a more general sense as fundamental a trait of the human mind as is that of the absorption and assimilation of ideas". Was there not originality in the invention of traps and snares, weapons, canoes, rafts, house and knots? So also who could doubt the independent invention of spirits, taboos, modes of navigation, methods of hunting, fishing, warfare, the making of fire, punishments, ceremonies, myths and social customs? Inevitably there occurred similarities, parallelisms and convergences between cultures in this respect.

Goldenweiser found ES's historiography to be deeply flawed. He was willing to name names. The great fathers of English anthropology –

Spencer, Tylor, Lubbock, Frazer, Lang – were not single-minded revisionists (as was implied by ES). They may have often abused concepts like "psychic unity" and independent origins, but they by no means totally neglected the concept of diffusion of culture. Nor did modern ethnologists embrace independent invention to the exclusion of ideas of contact of peoples and diffusion of culture. Goldenweiser worked with, and was a great admirer of, Franz Boas, also at Columbia, and was close to the eminent American anthropologists Robert H. Lowie and A. L. Kroeber, both at Berkeley. ES seemed to be caricaturing such scholars (and they were more than capable of hitting back). Goldenweiser placed ES in the diffusionist school that had arisen in Europe under the leadership of Friedrich Ratzel (who had had a considerable impact upon the young Franz Boas), but had gone to extremes under the influence of Fritz Graebner and Wilhelm Schmidt (we have seen that ES was acquainted with their ideas), and more recently Rivers. Goldenweiser found their method, as also that of the classical English anthropologists, flawed because they seized upon an explanatory principle and ruthlessly applied it in all instances. The method was "dogmatic and uncritical" (p. 532). Goldenweiser attacked Rivers a few months later,[9] and was to expand his case against the diffusionists in his influential textbook *Early Civilizations: An Introduction to Anthropology* (1922).

Other criticisms were more pragmatic. The Bostonian Philip Means regarded ES's American speculations as important, and supported by much corroborative evidence. But, he asked, if it were true that the American aborigines borrowed Egyptian pyramids, irrigation systems and many customs, why was there no such thing as a wheeled vehicle in all of pre-Columbian America (chariots were much used in ancient Egypt), nor any ships or vessels of advanced Egyptian types (even in civilizations such as those of Mexico, Yucatan and Peru)? It would be a long time, he predicted, before American anthropologists "will be forced to accept these views as final".[10]

ES hit back, in his usual brusque style.[11] He denied Goldenweisner's charge that he had embraced "the Graebnerian faith": "The writings of Graebner, Frobenius, Ankermann, Foy, Schmidt and Montandon were quite unknown to me when my conclusions were first formulated"; and in fact their views and his had nothing in common except that they repudiated an antiquated psychology. Indeed ES saw the Grabnerians as essentially in revolt against Darwinian evolution and against supposedly (but wrongly named) "evolutionary" ethnology. Diffusionism was in fact more consistent with Darwinism, "for it is a much closer approximation to the biological idea to look upon similar complex organizations of a series of artificial civilizations as having been derived from the same common source, just as all vertebrate animals were offspring of one stock, which, after spreading abroad, became more or less specialized in a distinctive way in each locality" (p. 243). What of the argument that similar needs and environments led to humans independently deriving identical cultural features? This, said ES, ignored the fact that "in the majority of cases such identities of culture actually occur under circumstances and to meet needs as dissimilar as they can possibly be" (p. 244). Again, kindred people living in very similar circumstances (as existed for example in the neighbouring islands of Indonesia or Melanesia) could produce a complex stone-using culture in one area and something completely different nearby. He responded to Means in similar vein. If there was any substance in the "psychic unity" hypothesis, why indeed had the civilized people of Peru and Mexico not invented more in the way of ship-building? Or why didn't the Americans "get a 'happy thought' and invent 'so simple and obvious a device' as a wheeled vehicle?" (p. 242). The sailors who carried Egyptian culture abroad travelled (obviously) by sea, and in any case at that time had quite likely never even seen, or heard of, a chariot (reserved for kings and potentates).

ES's book *The Evolution of the Dragon* appeared in 1919. As with many of his writings it was a loose compilation of recycled papers, talks and work-in-progress. His crowded life left him no time to write a more cohesive monograph. He covered topics ranging from the use of incense and libations to the symbolism of the dragon. ES had long complained that professional archaeologists and ethnologists – not to mention historians and other scholars – were too often immersed in their own local studies, too tunnel visioned. They did not read across disciplines enough, did not make comparative analyses, or raise their gaze to the larger picture. Thus they were not attuned to diffusionist ideas. The more he read in fields such as architectural history, comparative religion, ancient and maritime history, the more interconnections he saw. (Interestingly, criticisms are still being made that archaeologists and anthropologists are insufficiently interested in, or trained in, inter-disciplinary knowledge.)

As he explained in his chapter on "Incense and Libations",[12] ES could now relate his expertise on Egyptian mummification to developments in architecture, maritime trade and medicine. In fact the macro influence of mummification had been vital, not only upon Egyptian civilization but upon world history for many centuries after the decline of the Egyptians: "the practice of mummification was the woof around which the web of civilization was intimately intertwined". Mummification had shaped the innermost beliefs of humankind and directed the course of religion and science (including medicine). By giving shape and substance to the belief in a future life, mummification laid the foundation for a theory of the soul. It was the Egyptian embalmer who had taken the vague ideas of physiology and psychology that had developed since Paleolithic times in Europe, and crystallized them into a coherent form. It was the Egyptians who took pre-existing concepts of deities and gave them a much more concrete form. In fact it could be said that the practice of mummification had played a key part in establishing the foundations upon which all religious ritual was subsequently built. Thus, for example, a priesthood was initiated in order to administer the rites that were suggested by mummification. Religious "paraphernalia" such as "the swastika and the thunderbolt, dragons and demons, totemism and sky-world" were all connected with Egyptian mummification.[13]

The ritual use of libations and incense was linked to very early ideas of the life-giving power of water, developed probably soon after the invention of agriculture. The culture of pre-dynastic Egypt was, of course, based upon the extraordinary fertility of the Nile basin. The annual flooding of the Nile produced a wetland space bordered by desert. The Egyptians (ES thought) had from ancient times ingeniously devised irrigation systems that enabled cultivation of crops over a wider and wider area. As population grew denser, urban clusters grew up. In these clusters a complex culture eventually arose. Who became the power-mongers in such a culture? ES suggested they were the people, the elite, who controlled water, the Nile water system. This percolated into the whole myth system of the Egyptians.

People drew parallels between the procreating power of water and semen, and with concepts of a Mother Earth. In a key legend, fertilizing water came to be personified in the person of Osiris, whose consort Isis was identified with the earth which was fertilized by water. In the earliest pictures we have of rulers in Sumeria and Egypt, they were portrayed as makers of irrigation canals and controllers of water, givers of fertility and prosperity. Once concepts of an after-life arose, the belief also arose that the bodies of kings could not only be preserved but reanimated by the use of water as a libation: "it was natural that a wise ruler, who, when alive, had rendered conspicuous services should after death continue to be consulted. The fame of such a man would grow with age; his good deeds and his powers would become apotheosized; he would become an oracle whose advice might be sought and whose help be obtained in grave crises. In other words the dead king would be 'deified' . . . The dead king also became more real when he was represented by an actual embalmed body and a life-like statue, sitting in state upon his throne and holding in his hands the emblems of his high office" (pp. 30–31). From these early Egyptian attempts to unravel the mysteries of death could be traced a rich crop of world-wide myths about humans and animals dwelling in stones.

The Osiris legend had created a profoundly influential symbolism: "For Osiris was the prototype of all the gods; his ritual was the basis of all religious ceremonial; his priests who conducted the animating ceremonies were the pioneers of a long series of ministers who for more than fifty centuries, in spite of the endless variety of details of their ritual and character of their temples, have continued to perform ceremonies that have undergone remarkably little essential change" (p. 32). Ritual acts, such as the use of incense and libations, offerings of food and blood, baptism representing the giving of life, and so on, still persisted.

There were also vital links between mummification and architecture, an Egyptian art of the first order with immense world influence. How was this? The catalytic factor was the great importance attached by the Egyptians to the preservation of the dead and the making of provision for the dead in the afterlife. This led to the aggrandizement of the tomb. Stone-making became a skilled art as rock-cut graves and simple brick chapels evolved into the magnificent structures of later dynasties. As an architectural historian noted, stone buildings came to be associated with a sense of the sacred, "of magic stability and correspondence with the universe, and of perfection and proportion".[14] (Thus an hieratic tradition grew up that forbade the use of stone for civil purposes, a taboo that only broke down in the Roman empire.)

For over four thousand years Egypt invented new building methods and devices that were widely adopted by its neighbours. These included shaft-tombs and *mastabas* of the Pyramid Age which spread to the whole eastern Mediterranean, from Crete, Palestine, and Syria to southern Russia and the north African littoral. In ES's view these underwent various modifications in each place, and in turn became models that were roughly copied in later ages by wandering dolmen-builders. For instance: "The round tombs of Crete and Mycenae were clearly only local modifications of their square prototypes, the Egyptian Pyramids of the middle Kingdom" (p. 13). There was evidence that Aegean models had influenced northern and western Europe even during the Bronze Age. ES cited the chambered mounds of the Iberian peninsular and Brittany, New Grange in Ireland and Maes Howe in the Orkneys; and, further afield, "the influence of these Aegean modifications may possibly be seen in the Indian *stupas* and the *dagabas* of Ceylon, just as the stone stepped pyramids there reveal the effects of contact with the civilizations of Babylonia and Egypt" (p. 13). Egyptian influence was also evident in Christian churches (as in the domed roofs, iconography, symbolism and decoration of Byzantine architecture) and in Islamic buildings.

He began to concentrate on the diffusion of Egyptian culture to China. In this he was much influenced by J. J. M. de Groot's multivolumed work on *The Religious System of China* (1901). De Groot had closely scrutinized Chinese beliefs. ES drew the conclusion that – despite many obvious differences between the two cultures – they had much more in common than was usually thought. In particular he detected essential identities between Chinese and Egyptian concepts of the soul and its functions. The Chinese concept of the immaterial soul or *shen* was very like the Egyptian *ka* (as was the Persian *Fravashi*, and it was no coincidence that the Iranian domain lay on the overland route from Egypt to China). ES offered his "chain of proof" that Egypt was linked to China. The links in the chain were, in his words, as follows:

> (a) the intimate cultural contact between Egypt, Southern Arabia, Sumer, and Elam from a period at least as early as the First Egyptian Dynasty; (b) the diffusion of Sumerian and Elamite culture in very early times at least as far north as Russian Turkestan and as far east as Baluchistan; (c) at some later period the quest for gold, copper, turquoise, and jade led the Babylonians (and their neighbours) as far north as the Altai and as far east as Khotan and the Tarim Valley, where their pathways were blazed with the distinctive methods of cultivation and irrigation; (d) at some subsequent period there was an easterly diffusion of culture from Turkestan into the Shensi Province of China proper; and (e) at least as early as the seventh century BC there was also a spread of Western culture to China by sea. (p. 49)

In a chapter on "Dragons and Rain Gods" ES addressed the colossal subject of the dragon-legend in human history.[15] This was fertile but perilous territory. As he sought to find ancient Egyptian origins for the rich and diverse dragon myths that were flung across the globe, he was to become embroiled in perennial controversy. He took an almost obsessive interest in the subject, immersing himself in a massive literature. He saw the dragon legend as a complex, and in many ways confusing, expression of mankind's aspirations and fears over more than 5,000 years, as the myth evolved along with civilization itself. Essentially the dragon myth embodied the search for the elixir of life – the underlying force behind the fabric of civilization, preserved by popular tradition, constantly growing in complexity as it absorbed new events, forgot old meanings and invented new ones. It was a story of blendings, confusions and distortions, very similar to the dream-development of individuals but "vastly more complex than any dream, because mankind as a whole has taken a hand in the process of shaping it" (p. 77).

ES was well aware of dream analysis through the work of Freud and his friend Rivers (Rivers had given an influential lecture on "Dreams and Primitive Culture" in April 1918). ES was no Freudian. He seemed to have an emotional aversion to Freud's sex theories and he specifically rejected Freud's interpretation of myth (and was even more scathing about Freud's followers, notably Jung). Nevertheless he borrowed elements of Freud's dream theory. In a dream state (according to Freud) individuals no longer exercised a rational censorship over their stream of consciousness. All sorts of fancies were unleashed. A "fantastic mosaic" was built up out of aspirations, fears, and memories of a hodge-podge of events and experience. Myths (said ES) resembled dreams insofar as the controlling influence, or censorship, of a story-teller got lost as stories were handed down from generation to generation, and ultimately from century to century. Fortunately a vast mass of detailed evidence about the dragon legend had been preserved in the world's literature and folk-lore.

The dragon was intimately associated with the earliest Egyptian divinities, the earliest trinity of the Great Mother, the Water God and the Sun God:

> To add to the complexities of the story, the dragon-slayer is also represented by the same deities, either individually or collectively; and the weapon with which the hero slays the dragon is also homologous both with him and his victim, for it is animated by him who wields it, and its powers of destruction make it a symbol of the same power of evil which it itself destroys . . . In the saga of the Winged Disk, Horus assumed the form of the sun equipped with the wings of his own falcon and the fire-spitting uraeus serpents. Flying down from heaven in this form he was at the same time the god and the god's weapon. As a fiery bolt from heaven he slew the enemies of Re, who were now identified with his own personal foes, the followers of Set.

Such stories enabled people to identify with things around them, the skies, the weather and so on, but they also expressed the emotional conflicts of daily life, good versus evil, light and darkness, justice and injustice, wealth and poverty: "The whole gamut of human strivings and emotions was drawn into the legend until it became the great epic of the human spirit and the main theme that has appealed to the interests of all mankind in every age" (pp. 78, 80–81).

The dragon became a "composite wonder-beast" that ranged from western Europe to Asia and America. In the earliest depictions of a dragon (in Mesopotamia in Susa) it was compounded of the forepart of an eagle (Horus) and the hindpart of a lion (Hathor or Sekhet) but with human attributes and water-controlling powers that could be traced to Osiris (the fundamental element in the dragon's powers was control of water). In other times and places the dragon was made up of a veritable menagerie of creatures, ranging from stags, camels, or snakes to clams, carp and cows: " in most places where the dragon occurs the substratum of its anatomy consists of a serpent or a crocodile, usually with the scales of a fish for covering, and the feet and wings, and sometimes also the head, of an eagle, falcon, or hawk, and the forelimbs and sometimes the head of a lion. An association of anatomical features of so unnatural and arbitrary a nature can only mean that all dragons are the progeny of the same ultimate ancestors" (p. 81). The connection with water, and with Osiris, was another clinching factor in ES's argument. Dragons were almost everywhere water creatures: "It controls the rivers or seas, dwells in pools or wells, or in the clouds on the tops of mountains, regulates the tides, the flow of streams, or the rainfall, and is associated with thunder and lightning. Its home is a mansion at the bottom of the sea, where it guards vast treasures, usually pearls, but also gold and precious stones" (pp. 81–82).

For archaeologists seeking insight into dragon myths (and other myths) of the Old World, they had no further to look than to the Americas. The jetsam and flotsam of Eurasia had from earliest times swept on to the shores of the American continent, spanning the world almost from pole to pole:

> The original immigrants into America brought from North-Eastern Asia such cultural equipment as had reached the area east of the Yenesei at the time when Europe was in the Neolithic phase of culture. Then when ancient mariners began to coast along the Eastern Asiatic littoral and make their way to America by the Aleutian route there was a further infiltration of new ideas. But when more venturesome sailors began to navigate the open seas and exploit Polynesia, for centuries [circa 300 BC to 700 AD] there was a more or less constant influx of customs and beliefs, which were drawn from Egypt and Babylonia, from the Mediterranean and East Africa, from India and Indonesia, China and Japan, Cambodia and Oceania. One and the same fundamental idea, such as the attributes of the serpent as a water-god, reached America in an infinite variety of guises, Egyptian, Babylonian, Indian, Indonesian, Chinese and Japanese, and from this amazing jumble of confusion the local priesthood of Central America built up a system of beliefs which is distinctively American, though most of the ingredients and the principles of synthetic composition were borrowed from the Old World. (p. 87)

Thus America had more or less preserved in aspic every phase of the early history of the dragon-story. This was to be found in American pictures and legends, in an "amazing luxuriance" of symbolism.

ES ingeniously drew out parallels and resemblances between creatures and gods depicted in the rich folklore of Native Americans and those of, for example, India, China and Japan. Infuriatingly he used the evidence collected by American archaeologists – who were almost to a man anti-diffusionists – to weaken their position. To take an example, he cited work by the respected scholar Herbert Spinden on the Algonkian, Iroquois and Pueblo Indians. Spinden, whose book *Ancient Civilizations of Mexico and Central America* would come out in 1928, was a militant exponent of the view that New World civilization had independent origins. The two men were to spend considerable time sparring with each other in the years to come. ES – always ready for a fight but always ready also to give his opponents a platform – got Spinden to contribute to *Culture: The Diffusion Controversy*, edited by ES and published in 1927. Here Spinden flatly declared that America was "the home of a family of civilizations independent of the family of civilizations in the Old World".[16] He no doubt remembered that in 1919 ES had cited his work on the horned snakes in Native American folklore, then showed detailed resemblances with dragon legends in China, India and Babylonia. Spinden had compared myths relating to horned snakes in California to common legends in Algonkian and Iriquois folklore about monsters possessing antlers and sometimes wings. He noted that the horned serpent was usually a water spirit and an enemy of the thunder god (just as in Babylonian and Indian homologues, ES pointed out). Among the Pueblo Indians the horned snake was a spirit that lived in the water or sky and was connected with rain and lightning. ES was struck by the deer's antlers mentioned above. Dragons equipped with deer's antlers were a distinctive theme in Chinese dragon legends. Also snakes with less specialized horns described by Spinden suggested the Cerastes of Egypt and Babylonia.

To bolster his case ES reproduced a rock drawing in Piasa, Illinois, recorded by the missionary explorer Marquette in 1675, that seemed based on Algonkian dragon legends. The original rock work had long since disappeared but a surviving sketch depicted a winged serpent with deer's antlers, a body covered in scales and (in Marquette's words) "red eyes, a beard like a tiger, and a frightful expression of countenance. The face is something like that of a man, . . . and the tail so long that it passes entirely round the body, over the head, and between the legs, ending like that of a fish" (quoted p. 93, *Evolution of the Dragon*; also figure 3, p. 94). These anatomical peculiarities were arrestingly similar to those of Chinese and Japanese dragons and, said ES, "are so extraordinary that if Père Marquette's account is trustworthy there is no longer any room for doubt of the Chinese or Japanese derivation of this composite creature". If the account were not accepted, he added, "we will be driven, not only to attribute to the pious seventeenth century missionary serious dishonesty or culpable gullibility, but also to credit him with a remarkably precise knowledge of Mongolian archaeology" (p. 95).

ES spent a great deal of time on dragon myths in China, Japan and India. Ultimately, he felt, they could be traced back to Babylonia, "the great breeding place of dragons" (p. 102). For those who might object that this had little to do with Egypt, he had an answer. True, Egypt had little in the way of the characteristic dragon and dragon-story. But the essential ingredients out of which the monster and the legends were compounded had been preserved in Egypt, and perhaps in an earlier and purer form: "Hence, if Egypt does not provide dragons for us to dissect, it does supply us with the evidence without which the dragon's evolution would be quite unintelligible" (p. 104). The crucial evidence was to be found in the myths about the Great Mother, the Water God and the Warrior Sun God and in the three legends: The Destruction of Mankind, The Winged Disk, and the story of Horus versus Set (or Seth):

> Babylonian literature has shown us how this raw material was worked up into the definite and familiar story, as well as how the features of a variety of animals were blended to form the composite monster. India and Greece, as well as more distant parts of Africa, Europe, and Asia, and even America have preserved many details that have been lost in the real home of the monster. (p. 104)

Briefly, the recurring theme of ambivalence – the fluctuating between good and evil – within dragon lore is traced back to Egyptian lore. The Great Mother (Hathor) was not only the benevolent giver of life but also the controller of life and thus a death-dealer. In a significant tale, in the very remote past, the Great Mother was summoned to rejuvenate the ageing king. The only elixir known was human blood. So the "giver of life" had to slaughter humankind. She became the destroyer of mankind in her lioness avatar as Sekhet. Early depictions of the dragon had a hind-part of a lioness (the mother goddess) and a forepart of the sun-god's falcon. The sun-gods Horus and Marduk were givers of light, overpowering the forces of evil, but at the same time, as warrior-gods, forces of destruction: "The falcon of Horus thus became also a symbol of chaos, and as the thunder-bird became the most obtrusive feature in the weird anatomy of the composite Mesopotamian dragon and his more modern bird-footed brood, which ranges from Western Europe to the Far East and America" (p. 106). The water god Osiris (Horus's father) represented life-giving moisture, rain, the inundating Nile and streams of the earth, the sea, and waters of the sky. He was the bringer of the wind (whose life-giving breath could raise a king from the dead) and lord of the wine. But the water god (or gods) was also identified with the destructive powers of flood, sea and wind, killers of humans. Whereas Osiris was depicted as a man, his equivalent in Babylonia, Ea, was represented as a man wearing a fish-skin, or as a fish, or a hybrid monster with a fish body and tail (the prototype of the Indian *makara*, the father of dragons).

In one form or another, the story of mankind's destruction (or the aversion of it by means of sacrifice) crops up across the globe and goes back more than three thousand years. It is closely associated with the myths of Hathor and Horus. In an early version of the story the deity Re (or Ra, a god worshipped during the fourth and fifth Dynasties, pictured as a hawk-headed man with a sun-disk on his head) is faced by rebellious sons of Horus. He calls on Hathor, the mother goddess, to slay them. She slaughters humans and the land is flooded in blood. Re is aghast and saves a portion of mankind by tricking Hathor into drinking barley beer mixed with a red ochre coloured powder (in ES's medical opinion an opium derivative; the red powder is also a metaphor for the red waters of the Nile at flood-time). She becomes intoxicated and no longer recognizes mankind, which is thus saved. Many dragon stories have a similar theme of a dragon being vanquished by a sleep-producing drug. As said before, this story supposedly mythologised the attempts of an early king to save himself from being killed when his powers began to fade by getting the Great Mother to rejuvenate him through blood sacrifice. Blood was supposed to have great healing powers in Egyptian pharmacy. As time went on human sacrifice (which appears in so many cultures) was abandoned. Substitutes were adopted for human blood, such as animal blood or coloured liquid (or later wine as a symbol of blood). This substitution was not necessarily a matter of compassion for human sacrifices. Various animals were identified with the gods. This explains the puzzling fact that so many mummified animals were found alongside kings and nobles in tombs. As an instance, the cow was identified with the Great Mother. This was the source of the cow-cult that extended to Africa and India. Osiris, Isis and Set were identified with the pig, which was also closely associated with the Great Mother and the Water God. Blood from such animals was therefore seen as more potent than human blood used in sacrifices to the gods.

There were many and confusing versions of the tale of the destruction of mankind. In some Re himself is rejuvenated by Hathor. The Re story can be seen as the start of the belief in the sky-world or heaven. Hathor, originally only a cowrie-shell amulet to enhance vitality, became personified as a woman and identified with a cow and a moon (hence the nursery rhyme "the cow jumped over the moon"). Re became the supreme sky-deity as the sun, and Hathor, as the moon, became the lesser "eye" of the sky. As the Eye of Re she was identified with the fire-spitting uraeus-serpent which the king or god wore on his forehead: "She was both the moon and the fiery bolt which shot down from the sky to slay the enemies of Re" (p. 116). This story became merged with an older legend of a fight between Horus and Set, the rulers of the two kingdoms of Egypt. There also emerged stories about the "Evil Eye" and petrification of the enemies of the gods, stories that are to be found world-wide. The tale of the slaughter of mankind also subtly morphed into the pervasive legend of The Flood, with Re's boat becoming the ark.

Elements of the story appeared in the Winged Disk Saga. Horus, as king of Lower Egypt, was identified with a falcon. He entered the sun-god's boat (the "boat of the sky", the crescent moon), sailed up the Nile and then mounted up to heaven as a winged disk, i.e. the sun of Re equipped with his own falcon's wings. On assuming the form of the winged disk Horus added to his insignia two fire-spitting serpents to destroy Re's enemies (paralleling Hathor's powers, swooping down from the sky like a bolt of fire). Thunderbolt weapons become a feature of later dragon-slaying myths. ES explored the great variety of weapons that cropped up in such myths from Babylonia to India. They ranged from the Cyclops eye planted in the forehead shooting out fire bolts, the lightning of Ishtar, Zeus's thunderbolt, the lightning sword of a Swahili hero, and so on. He even found parallels between the Horus-Osiris stories and that of the god Soma featuring in the Indian Rig Veda. The connection of dragon legend with animals such as the deer, ram and pig are also explored. There were signs here that ES was becoming obsessed with what was a rich and fertile discourse. His enjoyment of the stories was palpable. He retold them with zest, lingering over the minutiae, ingeniously seeking analogies and parallels. As he took to the highways and byways, he sometimes got carried away and risked losing the overall thread of his "scientific" argument.

This criticism could certainly be leveled at the third and final chapter (or lecture[17]) in his book *The Evolution of the Dragon*. It dealt with myths concerning the birth of Aphrodite. The Greek goddess of love, she is seen to have a close connection with the primal myth of the Great Mother, who had a vital role in the evolution of the dragon myths. ES used evidence from Arthur Evans (the Oxford archaeologist who pioneered Minoan studies) to suggest that the origins of the Aphrodite legend might be traced back to the African Hathor, the most ancient of the historic Great Mothers. Classical scholars had been too Eurocentric to explore such matters.

Aphrodite (or Venus in the Roman version) emerged into life from a giant sea shell. ES was able to use the researches of the eminent conchologist, and fellow diffusionist, Wilfrid Jackson to show the vital links between goddess myths and shells. Jackson was curator of the geological collections at the Manchester Museum and, like ES, a regular speaker at the Manchester Literary and Philosophical Society. He and Will Perry were good friends. Jackson had recently published a book *Shells as Evidence of the Migration of Early Culture* (1917), a book that convinced the usually sceptical A. C. Haddon of the spread of Old World culture to the Americas.[18] Jackson's book, which had been prompted by Elliot Smith,[19] analysed the cultural use of shells. The cowry shell in particular, because of its physical resemblance to the female vulva (a fact that ES and others used ingenious circumlocutions to explain), was identified with the feminine principle, reproductive and life-giving, also attributes of the early goddesses. (Here we have a case of ES using Freudian symbolism, despite his aversion to key elements of Freudian analysis.[20]) Cowry shells were used in the same way as red ochre, placed in a grave to confer vitality on the dead. This showed the assimilation of the life-giving properties of shells, blood and blood substitutes, topics ES expatiated on at length. Shells were also worn by women on bracelets, necklaces and girdles to confer fertility and good luck. Fully developed shell cults sprang up in the Eastern Mediterranean very early on, and may have owed something to cultural contact with people living on the coasts of the Red Sea and Indian Ocean.[21]

Goddesses were associated with a wide gamut of objects and creatures: "Aphrodite was at one and the same time the personification of the cowry, the conch shell, the purple shell, the pearl, the lotus, and the lily, the mandrake and the bryony, the incense tree and the cedar, the octopus and the Argonaut, the pig, and the cow" (p. 165, *Evolution of the Dragon*). These connections were not always simple and involved confusions of meaning and fancied resemblances, issues that ES tried to clarify by systematic scrutiny of an impressive range of writings in many languages and across fields and regions. It was an arduous task and inevitably, given his time constraints, incomplete and sometimes repetitious (perhaps a result of a lecture approach – the points made were not always easy to a non-specialist audience and needed added emphasis).

The elixir of life story, for instance, was recycled in various forms, and he gave detailed treatment of topics such as the origin of clothing, pearls, the octopus, the swastika (a symbol he suggested derived from the octopus), the mother pot (deriving from womb symbolism), portal legends, mandrake, the measurement of time (stemming out of the periodic phases of the moon and womankind and the identification of the Great Mother with the moon), the pig, gold, the thunder-stone, the serpent and the lioness, and of course dragons.

Let us take one topic: calendars and astronomy. He was to write a good deal more on this subject in his later writings on Mesoamerica, where he detected significant parallels between Old and New World calendrical and astronomical systems, parallels that have been supported by recent research (see *Afterword*). ES linked such phenomena to the annual flooding of the Nile. This was a vital event for the farming of crops and those with knowledge of its timing and regularities – such as the astronomer-priests – wielded great power as a result. A later commentator has summarized the theory of ES and Will Perry about the role of the astronomer-priests, those with the knowledge about the Nile's dynamics:

> They believe . . . that such advance knowledge was based on the observations of the moon in connection with the elaboration of a lunar calendar. As the Nile flood appears regularly after an identical number of moon phases (i.e. every lunar year) it was thought that the moon was somehow in control of the life-giving waters of the Nile. Furthermore it was also possible to connect the cyclical changes of the moon with the monthly menstrual periods of women, thereby establishing a causal relationship between the Nile water, the moon, and the life-producing functions of womanhood . . . From this apparent demonstration of causal relationships, the existence of celestial influences over mundane affairs was demonstrated. And this deduction was strengthened when it was realized that year in, year out, the star Sirius appeared regularly in the East, exactly at the time when the Nile flood began in the Middle East. Professor Elliot Smith has pointed out that the coincidence of the rising of Sirius with the regular inundations of the Nile was believed by the Egyptians to have the relation of cause and effect. It played an essential part, not merely in corroborating the hypothesis of celestial control of human affairs, which the moon's cycles had previously suggested, but also in helping to build up a comprehensive theory of the regulation by the sky of all the really vital affairs on the earth. He maintains that this archaic form of astrology was destined to exert a far-reaching influence upon thought and speculation for the next sixty centuries.[22]

Let us return to dragons and serpents. The serpent was identified with the Great Mother's destructive aspects. This Egyptian symbolism was spread widely. Hence the snake came to be seen as the most primitive form of the wicked dragon, a symbol of the powers of evil. This was the source not only of Babylonian myths (e.g., of Tiamat the serpent) but of the Biblical story of the serpent in the Garden of Eden. Eve was the homologue both of the mother of mankind and also the tree of paradise. ES summarized as follows:

> The earliest form assumed by the powers of evil was the serpent or

> lion, because these death-dealing creatures were adopted as symbols of the Great Mother in her rôle as the Destroyer of Mankind. When Horus was differentiated from the Great Mother and became her *locum tenens*, his falcon (or eagle) was blended with Hathor's lioness to make the composite monster which is represented on Elamite and Babylonian monuments . . . But when the role of water as the instrument of destruction became prominent, Ea's antelope and fish were blended to make a monster, usually known as the 'goat-fish', which in India and elsewhere assumed a great variety of forms . . . The real dragon was created when all three larval types – serpent, eagle-lion, and antelope-fish – were blended to form a monster with bird's feet and wings, a lion's fore-limbs and head, the fish's scales, the antelope's horns, and a more or less serpentine form of trunk and tail, and sometimes also of head. Repeated substitution of parts of other animals, such as the spiral horn of Amen's ram, a deer's antlers, and the elephant's head, led to endless variation in the dragon's traits.[23] (*Evolution of the Dragon*, pp. 232–233)

The "essential unity" of the motives and incidents of such myths across peoples and ages was a strong token – not of any "independent origin" – but of their derivation from the same ultimate source. Thus he ended the book on a familiar note.

In the turbulence that followed the end of the Great War and the onset of the deadly influenza pandemic, there was only a muted response to ES's *Evolution of the Dragon*. The world had had enough of war dragons at least. The general feeling seemed to be one of awe at ES's gargantuan effort but disquiet about his methodology and sweeping conclusions. As a reviewer in the *Dial* (20 September, 1919) commented:

> While the Professor of Anatomy at Manchester has tracked the dragon with provocative enthusiasm, daring conjecture, and unremitting ingenuity, he has also out-anthropologized the most doctrinaire anthropology in support of his persuasion that religion had a single source, from which it spread over the earth, and he has permitted himself enough disorder, repetition, and discursiveness to weaken confidence in his methods. (p. 274)

An anonymous review in the *Athenaeum* accused ES (and Rivers) of wavering between Graebner's ethnological diffusionism and Freud's psycho-analytical sex theory. Much was said about the sexual symbolism of the cowry shell. W. M. F. Petrie was the only major scholar to put his name to a review, but it was fragmentary and non-committal.[24]

Alexander Goldenweiser renewed his attack upon ES and the diffusionists in his textbook *Early Civilization:An Introduction to Anthropology* (1922). His narrative has Ratzel, Graebner and Rivers as the key theorists of diffusionism, as against the formerly prevailing paradigm of the classical evolutionists. He respected Rivers's undoubted intellectual powers and contribution to ethnology (especially his use of more accurate and serviceable methods), but rejected the "high artificiality" of his central theory of diffusion (imaginative as it was). ES is depicted as a wild follower of Rivers (whereas in reality ES often took the lead in their close collaboration; Rivers was certainly the more rigorous theoretician of the two). Goldenweiser pulled no punches:

> ES has achieved the questionable distinction of outdoing the dogmatism of the evolutionist by his reckless utilization of diffusion as an interpretation of widespread cultural similarities, supporting his theory by a comparative material apparently as inexhaustible in quantity and handled as uncritically as was the comparative material of the evolutionist.

He had little sympathy for the hypothesis that a Megalithic culture originating in eighth century Egypt had spread through the Mediterranean, over southern Asia and the island expanses of Melanesia and Polynesia to the remote countries of Mexico and Peru:

> this idea, however alluring, would require a delicate technique and categorical demonstration before it could claim serious attention. The methods used by ES are, on the contrary, so loose that the entire speculative edifice erected by him can at best be regarded as another link in the chain of top-heavy hypotheses, born of uncontrolled flights of the imagination and unchecked by either patient research or a strict method of procedure.[25]

Goldenweiser was now willing to admit "the great importance of the diffusion of culture as a stimulant of civilization" (p. 324). Rivers in his *Melanesian Society* insisted that the contact of people and the blending of their cultures were the "chief stimuli" in the process that led to human progress. There was much to be said for this approach, said Goldenweiser. But he still held that "the diffusion of civilization from tribe to tribe is but one of the basic factors in cultural advance, the other factor being human creativeness, resulting in the independent origination of new things and ideas" (p. 324).

Goldenweiser asked some reasonable requirements from the diffusionists. The most obvious was that they treat the reception and assimilation of borrowed traits with subtlety, giving due attention to the cultural and environmental difference to be found in the receptor areas. He cited the very professional work of the Sinologist Berthold Laufer, who had traced the intricate history of the potter's wheel from ancient Egypt to much of the globe, including India and China. This contrasted with the more simplistic approach of some others. He obviously had ES in mind:

> Clearly, the conception that diffusion is a quasi-mechanical process of the physical transplantation of cultural traits from one tribe to another, cannot withstand serious criticism. It is not enough to realize that a cultural feature leaves its original home, travels and arrives in a foreign tribe. It is equally important to know how and why it departs, what fates befall it in its wanderings and what reception it receives in its new home. (p. 319)

As we have seen, ES had long claimed to be doing precisely this, always admitting, and studying where possible, the cultural adaptations of features borrowed from the Old World. He defended himself vigorously, but the charge that he was an uncritical mechanist was to haunt him for the rest of his life, and was still cropping up in the 1970s. The tone adopted by Goldenweiser became increasingly the tone of professional anthropologists.

CHAPTER

4

Elephants and Ethnologists (1920–1924)

By 1920 Elliot Smith had become a towering figure in British medical science. He was a world authority on the comparative anatomy of the brain. Not only had this led to his abiding interest in Egyptian archaeology but also human prehistory and evolution. He was made a fellow of the Royal Society as early as 1907 and presided over the anthropology section of the British Association in 1912, when Rivers famously put forward his ground-breaking paper. Just as important was ES's presidential address, in which "he dealt with the identification of the neopallium, the associational area in the frontal region of the brain, and demonstrated its bearing upon the evolutionary problem in man, views then made known to a wider public for the first time".[1] His research threw new light on the role of neurological factors in human evolution. These factors had shaped the development of stereoscopic vision, speech, form and function. His use of endocranial casts was a major step forward in the discourse of human palaeontology and theories of early man. Less fortunately for his ultimate reputation, ES played a major part in publicising the Piltdown Man "discovery" of 1912, a supposedly ancient protohuman fossil whose combination of large brain and ape-like traits was thought to supply the "missing link" in human evolution. Like other experts of the day ES was taken in by the forgery, which was not revealed until after his death.[2] He was active in analyzing some of the many hominid fossils that were being located from the 1920s onwards in Africa especially, including Raymond Dart's famous *Australopithecus* skull found in 1925. A member of most prestigious scientific bodies in Britain – he was during his career president of the Anatomical Society, prominent on the General Medical Council, vice-president of the Royal Society, fellow of the Royal Society of Surgeons, president of the Manchester Literary and Philosophical Society and influential in the Royal Anthropological Institute – he was one of the mandarins of the British science establishment.

The eminent zoologist Solly Zuckerman later recalled "the enormous influence that Elliot Smith wielded in his day. At a time when the experimental method in biology had been all but taken over by physiologists and biochemists, he dominated the world of anatomy, in the same way that Rutherford, his close friend, dominated the world of experimental physics. By a strange coincidence, both men had been born in the same year, both had come to this country from the Antipodes, and both died in the same year".[3]

In 1919 ES abandoned his beloved Manchester to take up the chair of anatomy at University College in the University of London. This was part of a grand strategy. He wanted to integrate sciences such as medicine with social sciences such as anthropology into an overarching science of society (a dream of the age). He was attracted to London because, with the crucial aid of the Rockefeller Foundation, University College had been targeted for major reforms in medical science. Plans for a major centre combining physiology with anatomy had been advanced as early as 1907 by the professor of physiology Edward Starling, but the world war blocked progress.[4] The Rockefeller Foundation saw University College as one major centre – the British Imperial centre – in its worldwide objective to improve the lot of humanity and further human progress by means of public health education combined with advanced research-based medical science. Two of their representatives – Wickliffe Rose and Richard M. Pearce – came to Europe in 1919 to look into the medical needs and methods of medical education, and were impressed by a system of clinical units being trialed in London. This was rather along American lines. Medical disciplines such as surgery and gynaecology were placed in units, headed by clinical professors attached to universities (rather than medical schools) who did research as well as teaching, and staffed by fulltime clinicians. ES was an enthusiast for a more modern system that would emphasize the role of research, in contrast to the older amateuristic system that, if anything, denigrated research.[5] ES was a friend of Edward Starling, a champion of the new system. Starling had urged ES to accept the offer of the anatomy chair at University College. ES also represented Manchester University on the Medical Council, which was closely coordinating with Rockefeller. It seems highly likely that ES personally encouraged Rose and Pearce to implement their plans for University College.

Rockefeller gave a huge grant in 1920 (over £1.2 million) that enabled University College to consolidate existing elements such as the medical college and hospital with the medical school. An impressive new building of Portland stone was erected in Gower Street to house the Institute of Anatomy, making it a world class centre for anatomical research. As *Nature* enthused, it gave tangible expression "to a wider conception of the scope of anatomy, which will now include histology, embryology . . . and neurology, the study of animal movements by cinematography, radiology, and anthropology, and in fact the study of man in the widest interpretation of the term, his evolution, structure, and the history of his movements".[6] ES had been appointed head of anatomy in 1919 and the remarkable inclusion of anthropology in the new institute which he now headed had his fingerprints all over it.

With Rockefeller money ES not only built up a cutting-edge anatomy centre, but also research-based anthropology, focusing on human evolution and the diffusion of culture. He recruited new people, with outstanding research talents, such as the anatomists Popa and Una Fielding, the forensics expert Bernard Spilsbury, and the great historian of science Charles Singer. He appointed Will Perry as reader in cultural anthropology in August 1923. Since 1919 Perry had been reader in comparative religion at Manchester University and had also given lectures on ethnology in the department of psychology. His big book *Children of the Sun* had just come out. ES had wanted Rivers for the job but Rivers refused to leave Cambridge.[7] Funding was provided for anatomical and anthropological research in India, Australia and China, and ES could claim some spectacular results. His pupil Raymond Dart discovered the famous (if disputed) Taung skull in South Africa in 1925. Another pupil Davidson Black excavated "Peking Man" in 1926. ES traveled to China to analyse the bones and the geological and topographical surrounds. He popularized the new palaeontology by lecturing widely on such discoveries and their implications for human evolution. However he was less successful in founding a dominant school of anthropology. As Graham Richards comments, "several factors undermined his aspirations to create a unified human biology programme ranging from sociology to anatomy: the death of Rivers (his closest ally), anthropologists' hostility to his extreme ideas on cultural diffusion, and the apparent failure to establish fruitful links with C. Spearman and J. C. Flügel, University College's psychologists".[8] Under the influence of Malinowski and new ideas such as functionalism, anthropology would move in other directions. ES ultimately lost a struggle for power and funding to Malinowski, who set up a more powerful anthropology school at the London School of Economics.[9] However this trend was to take many years (a fact often lost on commentators). Even at ES's death in 1937 Perry and others were still convinced that diffusionism had won the day.

The 1920s were a very busy time for ES. He had to set up his institute, supervising the new building, constantly sitting on committees, doing administration, attending conferences and lectures as well as carrying his normal teaching load and trying to continue his research. To give a flavour of his strenuous life, here is an extract from a letter of 17

December 1919 to his friend Warren Dawson, apologizing for not getting in touch soon after taking up his new London job:

> Since I arrived in London last August I have been so tied down by a multitude of varied duties that I have not had a moment to call my own. In fact I have not been able to cope with all the official duties I should have dealt with . . . in the coming vacation I have to sort my books and papers, prepare eight public lectures that I have to give in January and February, think out a new course of lectures on anatomy for Art Students, and write three articles to which I am committed!

After giving four series of public lectures in early 1920, he visited the US in April to tour leading medical schools as preparation for his organization of the institute, wrestled with "great accumulations of arrears" when he returned, then had "a real rest, revelling in the Story of the Flood, most of which I have now written up", then embarked on a fortnight's holiday to Leiden (letter to Dawson, 5 August 1920). Most of this "holiday" he spent in the Leiden Museum, and in lecturing at the universities of Utrecht and Groeningen for the Anglo-Batavian Society.

In 1921 ES produced the entry on anthropology for the new (twelfth) edition of the *Encyclopaedia Brittanica*. Large slabs of it were paraphrases of his publications, in effect a manifesto for diffusionism. He earned few friends among traditionalists by criticizing the previous contributor, the iconic Edward Burnett Tylor. ES dismissed Tylor's entry as too reliant on the "reactionary" school in ethnology: "the present tendency is to sweep away all such sophistry and introduce into ethnology the real scientific method".[10] Tylor had died in 1917 and was much revered as the father of scientific ethnology.[11] As ES wryly predicted he did not expect to be invited back. And so it turned out. Malinowski was asked to do the next entry, and in turn he politely turned the tables on the diffusionists.

A year later, on 4 June 1922, ES's close friend and collaborator W. H. R. Rivers died (tragically of a twisted bowel, a condition that would normally be rectified easily). ES wrote to a friend: "Rivers's death is a real catastrophe and compels one to make a new orientation in life, for it has upset so many plans that we were developing in common". Dawson recalled: "They were full of plans for the rejuvenation of anthropology and its bearing on the greater scheme to which he afterwards devoted much attention under the label of 'Human Biology'".[12] ES had access to Rivers's papers and spent much time over the next few years preparing some of his more important unpublished manuscripts for publication. As it turned out, this embroiled him in further controversy, as accusations surfaced that he had edited Rivers's work to favour his own diffusionist theories. ES was also involved in attempts to preserve and properly analyse (for example, by X-ray) the recently discovered Tutenkhamen mummy. Carter and Carnarvon's excavation fascinated the world and ES wrote and spoke much about the circumstances and significance of the finds. His newspaper articles were collected into a popular book published in 1923 entitled *Tutenkhamen and the Discovery of his Tomb*. In that year also he brought out a new revised edition of his *Ancient Egyptians*. This included some interesting personal reminiscences of his time in Egypt.

The year 1923 was an important year in other respects for the diffusionists. This was the year that Will Perry's best-selling book *Children of the Sun* appeared. The book was dedicated to ES and Perry gave warm acknowledgment of his debt both to ES and Rivers (of Rivers he said: "To him I owe everything as an anthropologist").[13] Although the book was legitimately seen as an expansion of the new historical theory associated with Rivers and ES, Perry was no slavish disciple. He aimed at a new global approach that would illustrate his method of "culture-sequences", and he struck out in many new directions. As he said: "I have followed my own line of thought, and have tried to produce a picture of some aspects of the rise and spread of early civilization as they appear to me" (p. 4). In some ways his comparisons of a range of civilizations and cultures is reminiscent of Arnold Toynbee's later study of civilization.

Denying the stage theory of social evolution, and also ideas of geographical determinism, Perry claimed to be using a primarily fact-based approach. What had to be ascertained was how humans behaved across a range of situations. This approach took into account individual drives (economic, sexual, etc.) – this won Malinowski's approval – but also the totality of society and social relations and functions. Facts would then lead to hypotheses. These would shed light on more theoretical matters of cause and effect. An hypothesis must stand or fall by the test of facts. Any evidence counter to it would be fatal to it. This, he claimed, was the merit of his method of "culture-sequences", which systematically compared earlier with later cultures:

> In many places comparisons may be instituted between two or more phases of culture. We may select one element belonging to the earlier phase, and inquire as to its presence or absence in the later phase. . . . If such comparisons can be carried out on a sufficiently large scale, some rough knowledge will be gained as to the manner in which culture has been modified all through the region in the course of ages. If it be found that, wherever the inquiry be made, the answer is invariably the same in respect of any cultural element,

> then the first step will have been made to the foundation of a stable theory of the history of civilization in the region. This method possesses the merit of being as 'fool-proof' as a method well can be: it is, moreover, capable of easy control, for the production of contrary instances will soon serve to jeopardize generalizations founded on surveys of culture-sequences. (p. 2)

He looked at major regions – the classical world, North America, Oceania, Indonesia, and India – and at a plethora of "cultural elements": from gold and pearls to stone implements, beliefs about the sky world and sky gods, great mother myths, mother-right, the underworld, dual organization, totemic clan system, to exogamy and givers of life. Conclusions were founded on each area of research. Only evidence passing the test was accepted (or so Perry claimed). The result was a tentative theory of the growth and spread of culture. Critics could (and did) point to gaps in the evidence or inadequate detail. Perry agreed that he had not covered the whole ground (who could?), and that his study of a major culture, such as India, was partial. But it was unreasonable to dismiss the whole project on such grounds: "All that is claimed is that the study of culture-sequences has invariably given the same results throughout the region; that, for instance, mother-right precedes father-right; sun-gods precede war-gods, and so on. The process may be likened to that of throwing a fine web of argument from one point to another, and then adding to those in position fresh strands derived from new inquiries, so as finally to produce a fabric capable of sustaining some weight, the amount borne depending on the number of strands" (p. 471). It is worth remembering these words of Perry, especially in view of the simplistic criticisms of his (and ES's) methods that were to be made in the 1970s.

Perry's book has all of the elements of the heliocentric edifice. The Children of the Sun were the kings of the Fifth Dynasty (about 2750 BC) and after, who identified themselves with Re, the great sun-god of Heliopolis, the seat of a powerful sun-cult. It became the capital of the dynasty, supplanting the older capital Memphis (p. 130, and chapter X generally). The Children of the Sun became a great ruling caste, they and their equivalents spreading heliocentric culture to nearby cultures and eventually across the globe. Their efforts resulted in the creation of a magnificent "Archaic Civilization". Other cultures borrowed from it, adding new intellectual dimensions as they translated the fact-based Egyptian culture into new terms as they came to grips with their own radically different environments. But the general tendency was for a loss of the original culture with time. Processes of degeneration set in. The diffusionists had little belief in the doctrine of progress that had been dominant in nineteenth century Europe, perhaps a response to the crisis of culture that beset Europe in the generation before the Great War, itself a major disillusioning factor.

Egypt alone had the favourable conditions needed for the creation of a uniquely advanced civilization. Agriculture was invented here as the Nile, flooding annually in desert surroundings, enabled people to irrigate. They learnt to expand areas of naturally occurring millet and barley by planting seeds in adjacent dry lands, now irrigated by the use of channels. (Perry borrowed here from the agronomist T. Cherry.) Hydraulics developed. The need to predict the annual rise of the Nile waters led to observations of the moon's movements and the invention of the lunar calendar. Those who controlled such vital knowledge became the first "vested interest" and attained positions of power. Complex sun and water cults developed, together with priestly castes and intricate burial rituals. The desert conditions favoured rites of mummification. Megalithic edifices, dual organization and the other features of archaic civilization appeared.

The research of the diffusionists concentrated on locating these clusters of traits, as we have seen. Perry added to it with detailed analysis of possible trade routes, locations of precious "life-giving" objects (pearls, gold, etc.), megaliths, myths and beliefs, culture patterns in areas influenced by archaic civilization and the history of change and degradation of the old culture. He endorsed ES's claim that independent invention could not explain the appearance elsewhere of clusters so uniquely Egyptian. Here is a typical example of this line of reasoning:

> In the face of the accumulation of the various cultural elements of the archaic civilization, a process vastly intricate in its workings, and often directly the result of certain natural features, such as the annual flood of the Nile, and of the ambitions of certain groups of the community, is it seriously argued that processes entirely different, that may have taken place in other parts of the earth, could have given rise to results that are so similar? In the whole region out of Egypt the rule was that of degradation of culture; in Egypt alone can growth be witnessed. Therefore Egypt must have been the originating place of the archaic civilization. Anyone who denies this will be faced with a question in probability that surely would daunt the most determined exponent of the doctrine of independent origin. He would have to explain how, in other countries, where the conditions were vastly different, men came to elaborate a heterogeneous collection of beliefs and practices entirely similar to those collected by the Egyptians; he would have to interpret all the stories of culture-heroes on an entirely different basis; he would have to deny the historical value of tradition. . . . (p. 472)

Finally, it is worth making the point that Perry was no uncritical admirer of archaic civilization. Indeed in many ways he preferred hunter-gatherer cultures for their (supposed) peacefulness and egalitarianism. He and ES both pushed the view that primordial humans and hunter-gatherers lived in a relative state of peace – almost a golden age – contrasted with what followed.[14] War was not an essential feature of human society but was the product of civilization. The Egyptians had much to answer for. The invention of agriculture ushered in an age of more complex social organization, competition for territory, rivalry of power elites and the class state. These were the causal factors producing the phenomenon of warfare, which "progressed" steadily throughout the ages with the spread of technology. Rivers and ES held that warfare had diffused from one or a small number of centres.[15] Perry also blamed the subordination of women on Egypt and the rise of a more militaristic culture: "It can be shown, beyond doubt, that the increasingly militaristic and violent spirit that has been acquired by man has resulted in the subjection of women. The sexes in the food-gathering communities are on a stage of perfect equality, the normal condition of things" (p. 493). Ethnologists now find such views about hunter-gatherer peacefulness to be problematic or downright wrong.[16] Perry himself was appalled by the horrors of the First World War, but he had a touching faith that the growth of liberal democracy would (or ought) to result in the steady adoption of less violent modes of behaviour.

Malinowski gave *Children of the Sun* a surprisingly favourable review. He spoke of its "astonishing erudition, width and daring of outlook, and a great ability for marshalling evidence", and said that it was bound "to leave a permanent mark on the development of anthropological doctrine". This was praise indeed from one who still described himself "as faithful to the old school". The work of Perry and ES, he judged, would force the old school to revise its position. If the new school embraced more psychological interpretations and deeper sociological analysis, "both are bound to find some common ground for collaboration, and there is some hope that a war of extinction may end in a true peace treaty".[17] Vere Gordon Childe, whose *Dawn of European Civilization* appeared in 1925, also approved of Perry's work. This was not surprising at one level as Childe took a cultural-historical approach to archaeology and recognized the significance of cultural migrations within early Europe.[18] A. C. Haddon was also supportive.[19]

Specialists continued to attack the Perry-Elliot Smith thesis. An expert on British megalithic tombs O. G. S. Crawford disagreed strongly with Perry's analysis. Perry, he said, had flawed knowledge of British archaeology and geology, confusing dolmens with barrows and making other mistakes. Crawford rejected the whole idea that Egyptians traveled to distant lands exploiting mineral wealth: "Where are these 'mining camps', these 'foci of cultural influence', these dead Egyptians buried abroad?" he asked, quoting from ES's revised edition of *Ancient Egyptians*: "One would imagine that there was some archaeological evidence, that the sites of some Egyptian mining camps had been excavated and proved to be Egyptian by archaeological methods, or that some graves similarly proved to contain Egyptians had been found 'abroad'. But that is not so. Where is the evidence that a single Egyptian ship left Egypt, or that a single Egyptian dies abroad on such a venture?" Crawford claimed that agriculture, ship-building, copper working and writing originated before the Egyptians. For his part Crawford favoured Babylonia as the fount of civilization.[20]

The eminent Egyptologist T. Eric Peet made a slashing assault on Perry and ES. He disputed in detail many of their accounts of Egyptian history and archaeology. As to Perry's *The Origin of Magic and Religion* (1923), an elaboration of religious themes in *Children of the Sun*, Peet concluded: "Mr. Perry has made a big incursion into Egyptology and he must therefore not be surprised if Egyptologists claim the right to test his evidence. I have done so, and I can only say that I find it wanting often on matters of fact and nearly always on exact method of presentation. Mr. Perry puts it through a process compared with which the Bed of Procrustes must have been comfortable".[21]

Across the Atlantic, Robert Lowie was equally dismissive. Lowie was a powerful figure in the anthropological establishment in America. Like Goldenweiser, Kroeber, Benedict and Mead, Lowie was a pupil of the great anthropologist Franz Boas, whose influence in America was paramount. Boas and ES were to collaborate against Nazi race theory during the 1930s, but in previous decades Boas exerted his influence decidedly against transoceanic or external sources for indigenous Mesoamerican civilizations. Boas believed that such civilizations were home-grown and that the duty of scholars was to focus almost exclusively on the internal dynamics and structures of these cultures. As has been said more recently: "The heavy concentration by Boasians and other Americanists on problems of the Western Hemisphere did lead to a certain parochialism. Man was seen as coming across the Bering Strait bringing relatively little in the way of cultural baggage. In the New World man experienced a number of radiations; as part of his adjustment to differing environments, he invented agriculture and eventually built civilizations . . . [The Boasian group] were left with no comprehensive explanation for the similarities of Old and New World cultures, unless accident could be called an explanation".[22] Any dissident theories from Americanists were stamped upon forcefully, so it is no surprise that ES and Perry were handed the same treatment.

Lowie and Kroeber were leading figures at Berkeley. Lowie attacked *Children of the Sun* as "parochially dogmatic", ethnographically inadequate, factually mistaken and not up with the latest anthropological thought: "It does not occur to Mr. Perry that before assuming an alien origin for Central American civilization it would be desirable to consider the possibility of its derivation from the ruder aboriginal cultures which are disdainfully neglected throughout under the not wholly appropriate name of 'food-gatherers'".[23]

Lowie was to write the section on "Social Anthropology" for the next edition of the *Encyclopedia Britannica*, which appeared in 1926 (Lowie probably wrote it around 1924 or 1925). In it he steered between the "extreme" schools of evolutionary stage theory (or "uniform sequences" as he called it) and diffusionism: "fatal defects of method makes acceptance of either scheme impossible . . . Historical realism will rigidly exclude as useless for comparison any but sharply individualized features; it will trace distributions over continuous areas, will appeal to the ascertained facts of linguistics and other sciences; and, above all, will use findings of archaeological stratigraphy as the surest point of departure and ultimate arbiter in questions of chronology". However he made the crucial admission that the facts of distribution did seem to show certain Asiatic influences upon Mesoamerica. For example: "The sinew-backed bow, which in Egypt goes back to the second millennium BC, is characteristic of many Asiatic peoples, and occurs in the New World only north of Mexico and west of the Missouri. It may, therefore, be interpreted as a comparatively recent acquisition from Asia that failed to establish itself beyond the far west of the United States". The cultivated banana was another interesting example. It was invariably propagated from sideshoots and had long ago become seedless, "hence its range over a wide area in southern Asia and Polynesia is incontrovertible evidence of diffusion". Study of other customs and myths threw up strong evidence of possible connections between Old and New Worlds (such as in linkages between Lapp tambourines in association with shamanistic office appearing in Siberia and North America; or again shared folklore relating to the Magic Flight story, found in Europe, Siberia and America).[24] Eleven years later, in 1937, Lowie was to make a scathing attack upon ES in his influential textbook *The History of Ethnological Theory*.[25]

Perry made another powerful enemy in Britain, John Linton Myres, who was to become perhaps the most potent figure in the anthropological establishment there. Myres, a classical archaeologist and geographer, was president of the Royal Anthropological Institute from 1928 to 1931, general secretary of the British Association from 1919 to 1932 and editor of the leading anthropological journal *Man* from 1931 to 1946. He and ES had crossed swords at the Sydney meeting of the BAAS in 1914 (as we have seen) and were to continue to do so across the twenties and early thirties. Myres prided himself on logical and methodological rigour, and found little of it in diffusionism. He was respectful of Rivers, and of ES (on occasions), but delighted in finding errors in their work. He gave a severe review of *Children of the Sun* in the spring of 1925, challenging the assumptions behind the "historical method" generally, and Perry's "sequence of culture" theory in particular.[26] He accused the historical school of falling short of the "standards of accuracy and cogency which historians, in the more special and popular sense, are wont to observe" (p. 13). He attacked the diffusionists' "peculiar and persistent misuse of negative evidence". They habitually argued "as though a blank in the record were positive evidence that there existed there something which excluded that which is being sought . . . or as though the absence of a belief or custom from a given region now was sufficient proof that it had formerly been present, but had afterwards disappeared . . . In the 'general theory' built upon this precarious foundation, it is no wonder if the hypothesis of degeneration is in frequent use" (pp. 19–20). Myres was unimpressed by Perry's "givers of life" hypothesis and his "disregard of chronology" (p. 24).

Perry took up the challenge, defending both his facts and theory. He was clearly beginning to suffer a siege mentality. There were "attacks on all sides" being made upon the claims advanced by ES and himself: "These attacks are of interest; for while they fail to touch our position, they reveal the frame of mind of the attackers". It was significant "that all our critics feel it necessary to misrepresent our claims before they begin to destroy the figment of their imagination. Professor Myres has carried this process perhaps further than any other critic, as there is scarcely a paragraph in his address that is free from grave misrepresentations".[27] Myres replied briefly, reasserting his criticisms, and commenting: "I am sorry if I have annoyed Mr Perry, as some of his phrases suggest; still more, if I have misrepresented his views. I can only assure him that if I did so, it was not willfully, but because, in choosing between statements inconsistent with each other, I sometimes selected as expressing his real views passages which actually contradicted them".[28] Relations between the two men became decidedly frosty.

Meantime, inevitably, ES had been busy. In February he gave the annual Galton Lecture to the Eugenics Society in London. This lecture was in memory of the founder of modern eugenics in Britain, Francis Galton. ES was not a eugenist, and later campaigned actively against the spread of Aryan racial doctrines.[29] However, like many scientists (including Perry), he agreed with some eugenist ideas, and was not averse to using the Eugenics Society as a platform for his own theory. Asked to talk on racial characteristics, he managed to give an exposition of his (by now well-known) diffusionist thesis. His main aim was to dissociate race and culture (a prevalent tendency that he regarded as unscientific). Historians and anthropologists (he claimed) constantly explained features of culture and civilization as inevitable expressions of racial characteristics. This had some affinities with his pet aversion, the idea of independent invention or spontaneous evolution of customs, this time a product of innate racial traits. As he said:

> The varying temperaments of different races are patent enough, and their influence upon the intellectual and moral aspects assumed by culture in different communities can be clearly demonstrated; but it has not been generally recognized how large a part has been played by the social environment created by historical circumstances in shaping customs and beliefs and in determining intellectual and social progress. Many travelers, like Galton himself, have been impressed by the high intelligence and ability of certain peoples of lowly culture, and have realized that only the lack of opportunities for profiting from what civilization has put at our disposal has prevented such people from attaining a cultural status such as we enjoy and suffer.

ES was not taking the racist position of assuming an inherent superiority of the ancient Egyptians (or Sumerians, or whomever) who first invented civilization. (He was to be accused of this in the 1970s and later.) The Egyptians had been lucky. They had enjoyed special geographical advantages and a special series of events. In this milieu they had devised their splendid culture, and spread it abroad; but others equally well-equipped mentally used and adapted archaic culture. Nor were the children of the sun or those who accepted their culture necessarily superior to, or "better equipped", than others who were never presented with this opportunity – because of geographical isolation, for example. The question of race, he concluded, played a comparatively insignificant part in the adoption of culture.[30]

Eugenics also played a part in another of ES's ventures. He was invited to lecture at the University of California in 1924. That led on to a trip to Australia that helped in the founding of anthropology as a viable discipline in Australian universities. In December of the previous year, 1923, the American Eugenics Research Association, an influential body in the American academic world and politically (a contrast to the British situation), requested funding from Rockefeller for a study of natural selection within the indigenous population in Australia. The Eugenics Research Association had been founded in 1913 under the auspices of leading eugenists like Charles Davenport and Harry H. Laughlin. Extreme racial theorists such as Madison Grant and Lothrop Stoddard were key leaders within the association. It was Madison Grant who forwarded the Australian project to the Rockefeller Foundation. His genocidal theories, as later expressed in *The Passing of the Great Race* (1936), were to influence Hitler (who was in fact to declare American eugenics an inspiration for Nazi Germany).[31] The motivation for Grant's project seems to have been to show that "natural elimination" of inferior races or groups, such as the Australian Aborigines, was the way to go, in contrast to European welfarist policies that artificially preserved "defective" people from the effects of natural selection.

ES was well-known to Rockefeller, of course, and he was asked for advice. Whilst in the US he consulted with Rockefeller people (such as John Embree) and also with Davenport (head of the Cold Spring Harbor eugenics research centre): "Somewhat to the consternation of the [Eugenics Association], which felt control of the project slipping away, Embree sent Elliot Smith on to Australia to investigate the local anthropological situation, following a general Rockefeller Foundation policy that its own initiatives must always be linked to some local commitment of resources and personnel".[32] Warren Dawson (who had access to ES's papers) claims that when in New York ES was consulted by the foundation regarding the establishment of a department of anthropology at the University of Sydney and to draw up a plan of campaign of field-work in connection with it. ES was in Australia from August to October. In September he had a meeting with the Prime Minister, Stanley Bruce, aimed at gaining government support for a Sydney chair in combination with Rockefeller funding for the promotion of field-work and research into aboriginal problems. ES also gained financial support from two state governments for anthropological teaching. The Sydney department was duly established.[33] ES and A. C. Haddon were two of the electors who chose A. R. Radcliffe-Browne for the chair (in late 1925). Radcliffe-Browne was to rival Malinowski as the leading figure in British (more strictly British Commonwealth) anthropology during the thirties.

So hectic was ES's life at this time that he was unable to collaborate fully with his young protégé, the Egyptologist Warren Dawson in a book they had planned as a follow-up to the excitement aroused by the Tutenkhamen findings. This was a detailed account of mummies based on ES's researches and papers. He urged Dawson to write the book himself from these materials but Dawson preferred to wait: "Eventually, however, he [ES] found time to write only the last two chapters of the book, *Egyptian Mummies*, which was published early in 1924".[34]

Much of ES's time was spent in preparing his book *Elephants and Ethnologists*, which also appeared in 1924. ES's aim was to present a detailed study that would support his thesis of foreign influence in Central

America. This was in order to counter criticism that he made sweeping claims without sufficient empirical backing. So he revived and expanded a controversy that he had first begun in 1915 in an examination of a Mayan artifact.[35] This was a huge monolith, or stela, located at Copan in Honduras, an important centre of Mayan culture, where there was an imposing collection of ruins of stone pyramids, terraces and walls enclosing great courtyards (together with hieroglyphics that have more recently been deciphered as dynastic sequences).[36] Stonemasons had carved in high relief a series of these large monoliths and erected them in the courtyards. The site was variously dated as first to fourth or sixth to ninth century AD (ES favoured the later dates). It had been excavated by the Columbian government in 1834 and by the Englishman A. P. Maudslay between 1881 and 1894. ES was struck by Maudslay's illustrations and description of one particular monolith, Stela B. To his mind the pre-columbian sculptor had carved the picture of "an unmistakable Indian elephant ridden by an equally characteristic *mahout*". [37] [See Figure 1 on following page, which is from p. 4 of *Elephants and Ethnologists*].

ES had chosen this case-study in 1915 for a number of reasons. It was, he recalled, at a time when he, Rivers and Perry had become convinced of the reality of the diffusion across the Pacific of "the essential elements" out of which Mesoamerican civilization had been built: "But we were met by a solid phalanx of opposition to this interpretation" (p. 2).So he focused on one sharply defined issue, but one with far-reaching implications. The Copan elephant seemed so clear-cut that it would confound even the most ingenious believer in independent invention. How on earth could they argue that a people who had never seen an elephant could "independently" create the profile of an Indian elephant and a turbaned rider? Moreover by removing the "Egyptian factor" from the equation in this specific case he would remove "a disturbing element that had come to be so obtrusive as to hamper serious argument" (p. 5). (He believed of course that there was a Middle Eastern influence upon India, but he had already made that general case.)

However he had underestimated his critics. In 1924 he called them upholders of "the ethnological Monroe Doctrine". This was the nationalistic doctrine "that no outside influence could be admitted as a factor in America's cultural development". They were (he said) acutely aware "that the recognition of any Asiatic designs would be fatal to their belief in indigenous origin" (p. 5). They needed to explain away what common sense said was plainly an elephant in Copan. Thus they variously described it as a tapir (Maudsley's original suggestion), a tortoise, a Blue Macaw or similar parrot, or some vaguely remembered image of an extinct Pleistocene elephant or mammoth. Systematically ES set out to demolish such arguments. Onlookers sneered at the sight of learned men

Figure 1 Stela B at Copan, based on drawing by A. P. Maudsley
From page 4, *Elephants and Ethnologists* (1924)

disputing endlessly over macaws and elephants, but to the participants a vital principle was at stake.

ES drew on his own anatomical expertise, and on the supporting evidence of zoologists, to demonstrate that the sculpture was indeed that of an Indian elephant. Any errors in the analogy were seen to be due to the sculptor copying from representations of an elephant. The spiral ornamentations that adorned the elephant were shown to be remarkably similar to arbitrary conventions found in ancient images of the elephant, and fabled "elephant-fish" (makars), in India and Indo-China. The Copan stelae were erected in close association with "great pyramidal structures of a very distinctive type", very similar to Mesopotamian pyramids of 2400 BC, the idea of which had been exported to Eastern Asia. ES found "exact parallels" to the elephant structures of Copan in India "where also occur the practices of carving models of elephants and the wearing of turbans as head-dress . . . the intimate association of pyramids with stone stelae embellished with sculptured representations of gods is also found in India as another element in the same complex, which forces us to recognize in Copan the obtrusive influence of Hindu culture" (pp. 28–29). Also the design on the stelae at Copan was clearly inspired by Indian prototypes: "the beliefs associated with the elephant-headed god in Mexico and Central America, which are depicted on the most ancient American manuscripts, are identical with those which the people of India have associated with the god Indra ever since the time of their earliest writings, the Rig Veda; and the god Indra was intimately associated with the elephant" (p. 30). Supporting evidence also came from Mayan manuscripts, which graphically depicted scenes taken from late Vedic myths (p. 48).

ES reasserted Arnold and Frost's view, expressed in *The American Egypt* (1909), that the stela probably had an ancient Indo-Chinese origin. This was indicated by the Indian elephant and turban head-dress, combined with the Chinese cast of features and ornamentation on the figures in the American monument.[38] ES also detected remarkable analogies in overall composition between the Copan monument [Plate 4, p. 23 of *Elephants and Ethnologists*] and ancient Indian sculptures, such as one found in Mysore [Plate 8, *ibid.*, p. 28].

ES reviewed in detail American and Indian legends associated with elephant-headed gods. (Worship of the elephant-headed god Ganesh was, and is, widespread in India.) ES concluded that the parallels were remarkable and called for much fuller investigation by ethnologists. His case was documented with over 50 lavish illustrations taken from America, India, Indochina, Cambodia, Indonesia and China to ancient Greece and Babylonia, even Scotland and England. These ranged from elephants to makaras (mythical crocodile-serpents), antelope-fish, spiral ram's horns, serpent head-dresses, celestial dogs, macaws, mandrakes, and lion couches to wing discs and, of course, dragons.

ES's comparisons went well beyond elephants. For instance, he pointed to lion-couches. Fantastic lion-couches had been found in Tutankhamen's tomb. These were symbolically significant. They were vehicles that conveyed the dead king to the celestial regions and to immortality. Thus the lion-headed couch came to be seen as "the common determinative of a god" (p. 94). The idea was exported to both India and America. ES gave examples. One of his plates showed a two-headed lion couch bearing a Mayan god sitting in an Indian attitude on a stool. He was wearing a Phyrigian cap embellished on the side with a spiral Amen-horn. ES found parallels not only with many Indian drawings but with "a curule chair depicted on a Consular diptych of Flavius Clementinus (who was consul at Constantinople in 513 AD) . . . [this drawing] illustrates so strikingly the manifold cultural influences of the Old World that became blended in one object in the New" (pp. 94–95). ES also found homologies between the animal vehicles and weapons used by the gods depicted in both India and America.

From his case-study ES concluded that "the chief source of the inspiration of Maya art can be located in Indo-China [and that] the period that exerted the main influence was from the fifth to the twelfth centuries AD" (p. 127). The lines of diffusion (as he boldly depicted them in a chart p. 3) were a main stream from Cambodia (itself the recipient of Indian influence) to Micronesia and thence eastward across the Pacific to Central America. Another current went via Melanesia and a more southerly Polynesian route to South America. ES's thesis depended upon an earlier date for the occupation of Polynesia than the conventional seventh century AD. Rivers had argued for the early peopling of the Pacific for some time (for example, in his *Psychology and Politics* and his *Psychology and Ethnology*) and there was massive evidence appearing on the maritime activities of Pacific peoples and their capacities for long-distance travel.[39] Whether they had got to America was still problematic, but ES was confident that they had. And among the artifacts that they probably brought with them were small model elephants carved in sandal wood. Mention of them had been made in Chinese references to Cambodia in the early centuries of the Christian era: "Such trinkets carried by seamen might have provided the model which enabled the Copan sculptor to reproduce the elephant's profile with such remarkable accuracy" (p. 118).

If ES thought that he had settled the issue, he was soon disillusioned. Ethnologists disputed his readings of the Copan artifact and adjectives such as "extravagant", "sweeping" and "extreme" were applied to his thesis. They were to become embedded in the historiography. An anonymous reviewer in *Nature* was unpersuaded that the "zoomorphic designs" on the stela depicted an elephant. He could not discern a "tusk" or an "elephant goad", while the attitude of the supposed mahout seemed not to be that usually assumed. He suggested that the designs on the stela were symbolic and decorative and had been arrived at "by a fusion of prototypes. Earlier representations of macaws, cephalopods, tapirs, or other indigenous animals may well have contributed in the evolution of this complex hybrid design". He called for a more intensive comparative study of the art of the Mayas (so had ES), concluding: "In the long run, possibly, Prof. Elliot Smith's hypotheses may prove to be justified, but that stage has by no means yet been reached, and, in spite of his confident and over-emphatic assertions, his case cannot be accepted as proven".[40] Other reviewers made similar remarks, while whole swathes of evidence put forward in the book were simply ignored.[41]

As ES had wanly complained in his 1924 Galton lecture, life was not easy for those (like himself) who broke down barriers. They tended to be attacked by those on both sides of the barriers. There was certainly a perceptible undercurrent in ethnological comment on ES that it was ethnologists who knew best, and that even distinguished outsiders trespassed upon their territory at their peril. Anthropologists I know tell me that little has changed!

Ethnologists of the time claimed of course that they were being purely objective in resisting diffusionist ideas. However ES's warning about the role of nationalistic values in ethnological attitudes must resonate with historians. As every student who has done historiography knows (or should know), the rise of nationalistic movements in late nineteenth and early twentieth century Europe and Anglo-America significantly influenced national histories, most blatantly in Wilhelmine Germany but undeniably also in the United States, Britain and Australia. The same is true of sub-disciplines such as art history. Art historians have pointed to patriotic influences upon art movements, such as Impressionism in Europe and, quite significantly, in America. It would be surprising if American ethnology were immune to such cultural impacts. Even today, when American anthropologists and archaeologists have been forced to concede at least a degree of "foreign" intrusion upon indigenous culture (just as they have vastly increased knowledge of the indigenous roots of such culture and intra-American diffusions), many still seem emotionally resistant to evidence of inter-societal contacts before European colonization.[42]

ES now turned his attention to human prehistory and evolution. These were rapidly advancing fields and he wanted to synthesise existing knowledge for a wider public audience. He did so in 1924 with the publication of his book *The Evolution of Man*. The book enabled him to bring together a number of his own papers and writings, while at the same time providing what he believed was essential context for his ideas on the diffusion of culture.

CHAPTER
5

The Evolution of Man (1924–1927)

ES's diffusion theory did not come out of the blue but was related to his wider evolutionary views. These are set out in his *The Evolution of Man* (1924), but date at least from his 1912 address to the British Association (which is printed as chapter I). In compiling *The Evolution of Man*, his mind was concentrated by a perceived need to correct the thesis argued by his former assistant Dr. F. Wood Jones, in his books *Arboreal Man* (1916) and *The Problem of Man's Ancestry* (1918). In these works Wood Jones contended that the human family derived directly from a Tarsius-like animal rather than from apes and monkeys. This claim had sparked quite a controversy. Wood Jones pushed his theory in public but had been severely put down at the Zoological Society in 1919 by ES, R. I. Pocock, Peter Chalmers Mitchell and others. ES reiterated the attack in his book, putting recent evidence to reassert humanity's closer kinship with apes. This fitted in neatly with ES's overall argument that human evolution was not the product of separate bursts of independent invention but of the spread of culture and materials.

According to ES's pedigree, tree shrews and lemurs were indeed significant. By living in trees the shrews enhanced their sense of vision, touch, agility and hearing (at the expense of smell), factors that led to brain development, ultimately ushering in the order of primates. By the early Tertiary Period primitive lemurs had split into lemurs and tarsiers. The latter enhanced their power of stereoscopic vision and were ultimately transformed into early monkeys, brainier and more skilled in movement. This is said to have happened somewhere in Central America. At some stage the three branches of primates – lemurs, tarsiers and monkeys – migrated from the new to the old world via the land bridges that then connected the continents. During eons of time and vast wanderings, new types developed, including catarrhine monkeys and anthropoid apes, the latter wandering far and wide in Africa, Europe and Asia, increasing in size and adaptive power. They were the ancestors of the giant apes and early humans. While many types of lemurs, monkeys and apes became specialized, those in "the main stream of development that leads straight to Man" retained a primitive structure and plasticity that was the key to human adaptation and versatility.[1] (ES of course shared the anthropocentric values of his time, the idea that everything led to man.)

The higher mammals possessed higher brain power, notably through the development of the cortical area that ES dubbed the "neopallium", a coordinating centre for a complex system of visual, acoustic, tactile and memory-recording impressions. Superior memory enabled mammals better to learn from experience and to adapt more effectively to their environment. They thus evolved behaviours that began to transcend the more purely instinctive behaviours of creatures further down the evolutionary scale. As this new breed of intelligent creatures spread throughout the world, their adaptive powers resulted in "a bewildering variety of specializations in structure, some for living on the earth or burrowing in it, others for living in trees or even in flight; others again for an aquatic existence" (p. 28). However with specialization came a penalty: they became committed to one kind of life and sacrificed plasticity of structure and thus ultimately adaptability. But the humble stocks that eventually led to hominids retained the supreme virtue of plasticity. This proposition was to become orthodoxy in twentieth-century biology. Biologists like Julian Huxley hammered the theme that humans were distinguished from all else by vastly superior intelligence and plasticity.

ES argued that it was the steady growth and specialization of the brain that was the basic precondition for human evolution. He disagreed with other explanations that focused on factors such as the adoption of an erect posture or acquisition of speech, significant as those changes were. By themselves they were not enough. Some early gibbons, for example, could walk upright, thus freeing their hands, but they never developed advanced manual skills as they lacked the complex brain structures required. With early human-like creatures a sort of reinforcing cycle took place. Better brains enabled more sophisticated manual actions. These actions in turn enriched the brain through new elements of experience and knowledge gained. Consciousness, "meaning" and understanding of causes and effects gradually arose, and so did the faculty of shaping conduct in anticipation of results. ES, like Darwin, favoured an "out of Africa" thesis. He visualized early ape-men leaving their arboreal life in the African forests to seek new sources of food and opportunities, meeting new environmental challenges and, through selection, ultimately changing forms into new types.

No longer tied down, like the apes, to forests and tropical temperatures, the human family wandered far and wide, into every region of the earth. This mobility was the key to ES's diffusionism. "It is important not to forget", he said in an address to the British Academy in 1916 (which became chapter II), "that Man has been a wanderer ever since he came into existence, and that a diffusion of culture has been effected by this means ever since he set out from his original tropical home" (p. 63).

ES was at pains to point out that, while primal instincts and emotions remained an important part of human makeup, they were not the basis of invention and artifacts. The latter were the product of culture, built up essentially from the innovation of speech, which facilitated communication and allowed the devising of symbols and group memory. Even in the simian stage, our ancestors had all the specialized muscles needed for articulate speech and "the cerebral apparatus" for controlling their movements and acquiring skills:

> All that was needed to put this complicated machinery to the new purpose was Man's enhanced powers of discrimination to appreciate the usefulness of communicating more intimately with his fellows and to devise the necessary symbolism . . . I think it not unlikely that the acquisition of such fuller means of communication with his fellows by vocal symbols may have been one of the essential factors in converting Man's ultimate simian ancestor into a real man. (pp. 64–65)

The hallmark of humankind was enhanced power of conveying information and the capacity to hand the accumulated experience of one generation to another. But this build-up of tradition and beliefs, "this almost wholly artificial intellectual and moral atmosphere" that coloured man's social outlook, was slow and difficult, with innovation the exception rather than the norm, the preserve of a few elect individuals. What ES was saying was that humans had inherited more or less general instincts and aptitudes. At no stage had they inherited "highly complex and specialized instincts" which impelled them "without any prompting from other peoples, to build megalithic monuments or to invent the story of the deluge, independently of other people who do the same arbitrary things" (p. 66).

As we have seen, ES blamed the "stage theory" of cultural evolution, long popular with anthropologists, for driving people into the camp of "independent invention" of culture. It was the faulty slab on which the whole edifice rested. ES was a vehement critic of the use of terms – such as "Palaeolithic" and "Neolithic" (coined by Sir John Lubbock back in the 1860s) – that suggested humanity had gone through certain fixed and inexorable phases of evolution. The facts "reveal how devoid of foundation is the mis-named 'evolutionary' theory that claims all these phases of culture as so many natural stages through which every people has passed in virtue of the operation of the blind forces of an arbitrary and inevitable process of evolution" (p. 92). The reality was vastly more complex, even chaotic, with flux and overlaps of time periods and industries, and confusing exceptions to the so-called rule.

Again as we have seen, ES had been profoundly influenced by the psychologist-ethnologist W. H. R. Rivers. In his epoch-making 1911 address to the British Association, "The Ethnological Analysis of Culture",[2] Rivers had told how his researches in Melanesia had converted him away from the current theory. In his study of custom and institutions he had started out, conventionally, from the assumption that "where similarities are found in different parts of the world they are due to independent origin and development, which in turn is ascribed to the fundamental similarity of the workings of the human mind all over the world". Given similar conditions, similar customs and institutions came about and then evolved along the same lines. Rivers found that this simply did not work as an adequate explanation of the facts. He was forced to admit that he "had ignored considerations arising from racial mixture and the blending of cultures" (quoted, p. 18). Rivers's "lucid and compelling account" distilled ES's own feelings and "first brought home to me the fact, which I had not clearly realized until then, that in my own experience, working in a very different domain of anthropology on the other side of the world [Egypt], I had passed through phases precisely analogous to those described so graphically by Dr. Rivers" (p. 18). His study of Nile Valley culture had also suggested the vital role of migration and cultural blending. He was reinforced in such views by the work of William Sollas who, in his *Ancient Hunters* (1911), traced the origins of distinctive customs, beliefs and art forms to the movement of early humans out of Africa and the middle east (the "cradle" of the race) to other continents.

ES gave a name to this way of explaining culture. He called it the "historical" attitude, which was based upon "the solid foundation of the known facts of history and human behaviour" (p. 111). To take an example:

> The American Indian's belief that a Dragon equipped with wings and Deer's antlers is a power controlling water is assumed to have been derived from Asia, where the same complexly-eccentric belief is entertained. Even though no official records have been preserved of the flight of this Asiatic wonder-beast across the Pacific Ocean, the 'historical' school of ethnologists is convinced that it got to America in very much the same way as the Spaniards' guns or the Englishman's steam engines. (p. 112)

ES contrasted the historical school with the misnamed "evolutionary" school that interpreted culture "by means of hazardous and mistaken analogies with biological evolution" (p. 134). This implied an instinct theory (he blamed the Oxford anthropologist E. B. Tylor for initiating this mode of thought; Tylor had instituted anthropology at Oxford in 1875). Such ethnologists could only explain the winged dragon case as independent inventions flowing out of "a highly specialized human instinct to dream Dragons. The only other possible escape is to drop all this puerile speculation and admit the patent fact that the American Dragon came from Asia" (p. 113). ES felt strongly that this divergence of opinion between the two schools had "sterilized a vast amount of laborious research" (p. 113). Such dismissive language about rival methodologies was typical.

ES supported the thesis that had been put by his friend Will Perry, most recently in his *The Growth of Civilization* (1924), that early man had been essentially peaceful. (Sir Arthur Evans's assertion, against some contrary evidence it must be said, that Minoan civilization was notably pacific also contributed to this widespread belief.) The evidence suggested to ES that "until the invention of the methods of agriculture and irrigation on the large scale practiced in Egypt and Babylonia the world really enjoyed some such Golden Age of peace as Hesiod has described. Man was not driven into warfare by his instinct of pugnacity, but by the greed for wealth and power which the development of civilization itself was responsible for creating" (p. 131). Irrigation works had to be undertaken on the Nile and Euphrates and this prompted organization and exploitation of labour for vast works of aggrandizement such as the pyramids. Warfare came about when attempts were made to enslave neighbours, with headhunting, etc. "and the motive that prompted it was the common human desire to secure ease and luxury, both in this world and the next, at the expense of one's fellow men. The exploits of this military aristocracy, the 'children of the sun', during the last forty centuries make up the greater part of what usually passes for 'history'. They have been so widespread as to have misled most sociologists into the belief that warfare was a manifestation of the primitive instinct of pugnacity, instead of merely a by-product of civilization itself . . . Such theories of 'survival of the fittest' are as inappropriate as Bernhardi's misuse of this biological phase" (p. 132). ES was referring here to General Friedrich von Bernhardi, the German military theorist, whose *Germany and the Next War* (1912) glorified war as a "biological necessity". Such ideas were widely held to have set the scene for the First World War.

Overall ES emphasized cultural evolution rather than innate factors. The intellectual and moral outlook of societies was explained by their history, not by "some blind mechanically working force of evolution". Human attitudes had been formed not by alteration of genetic mental structures but by cultural influences imposed by communities: "Whatever the inborn mental and moral aptitudes of any individual, whatever his race or antecedents, it is safe to say that if he were born and brought up in a vicious society he would have learned, not merely to converse in the language distinctive to that particular group of people, but in all probability to practice vicious habits" (p. 133). The driving force of history has been the moulding force of society and culture rather than individuals working things out from scratch, the gradual accumulation of arts and crafts and customs and beliefs, providing individuals with a ready-made supply of opinions and ways of acting, usually accepted without questioning. There was no general tendency to strive for what Europeans called "progress": "progressive societies are rare because it requires a very complex series of factors to compel men to embark upon the hazardous process of striving after such artificial advancement" (p. 134). Hence the vital factor of diffusion of culture.

The period around 1926–27 was a crucial one for the diffusionists, in retrospect the beginning of a slow decline in their influence. A few years earlier things had been looking bright. As we have seen, ES had built up an impressive Institute of Anatomy at University College. Will Perry was appointed reader in cultural anthropology in 1923 and enthusiastically set out to build up an anthropology school. He seemed on top of the world. His magnum opus *Children of the Sun* had been published and certainly made a mark. However the critics were hurtful. Perry seemed to suffer psychologically, feeling besieged, his scholarship questioned and sometimes his responses verged upon the mildly paranoid. To make things worse, he began to show symptoms of Parkinson's disease, which quickly worsened. He became dependent upon carers for his physical needs. Despite this he carried on his career for another fifteen years, showing amazing industry and undiminished intellectual capacity. Books and papers continued to flow from his pen. He ultimately took early retirement in 1939. A vigorous controversialist, he was remembered by colleagues such as Daryll Forde as a man of passionate convictions but also of warm humanity.[3]

Although the Rockefeller Foundation was initially keen to fund ES's ambitious plan for a scientifically-based anthropology, their attention soon shifted to a similar project at LSE under Malinowski. In the funding war that erupted between Malinowski and ES, both warriors in academic politics, Malinowski proved to be the more adept. He had almost taken a university post in his native Poland in 1923 before accepting a lectureship at LSE. Had he done so, ES might conceivably have prevailed. Malinowski's appointment was stage-managed by his mentor Charles Seligman, professor of ethnology, in a move directed against University College and ES.[4] Malinowski's promotion a year later to reader in social anthropology (a term he was to make dominant in the discipline) was made with Rockefeller funding. By 1927 LSE was getting major Rockefeller grants and Malinowski was appointed to the new chair in social anthropology. His successful strategy was to marry the idea of scientizing anthropology with plans to apply "functional" anthropology to practical problems of colonial administration, especially in Africa. This had the support of the British Colonial Office and fitted in with a Rockefeller objective of improving race relations globally.

The year 1926 was a key turning point. ES had applied to Rockefeller for a grant of £300,000 to be split between anthropology and psychology. Key Rockefeller officials, including Beardsley Ruml, director of the Laura Spelman Rockefeller Memorial, debated the proposal: "It was decided to reject the proposal and therefore to reject Diffusionist Studies. In response, Elliot Smith wrote a scathing letter, which included the comment that he found amazing ' . . . the action of the Laura Spelman Trust in subsidizing those who are wrecking anthropological study in this country'. Someone wrote in pencil at the side of the letter, 'This means Malinowski of London School of E'".[5] ES later complained that the Rockefeller Foundation had "torpedoed" his plans.[6] A funding proposal from LSE was successful in 1927. This was to become the pattern of the future. ES fought against it with all of his considerable powers and influence until the end of his life. But Malinowski proselytized energetically in Britain and America. His functionalism came to be seen as the new paradigm. He gathered a string of able people around him, and his students, members of his famous seminars – such as Raymond Firth and E. E. Evans-Pritchard – came to dominate the profession in Britain.[7] However schisms developed within functionalism during the 1930s, with A. R. Radcliffe-Brown's alternative scheme of structuralist functionalism challenging Malinowski: "some of Malinowski's more important students shifted their theoretical allegiance to Radcliffe-Browne, who after two decades of academic wanderings (from Cape Town to Sydney to Chicago), finally succeeded in 1937 to the chair at Oxford". Thereafter "it was Radcliffe-Browne's synchronic natural scientific study of 'social systems' – overlaid upon the Malinowskian fieldwork tradition – that gave British social anthropology its distinctive character".[8] E. E. Evans-Pritchard later commented: "It has been a misfortune for anthropology that it lost its time perspective in this country, largely under the influence of Radcliffe-Brown and Malinowski, the first with no interest in particular events and the second having an interest in only one immediate set of them (the Trobriand Islanders); neither being really concerned with history of any kind or form".[9]

One of ES's more satisfying achievements in 1926 was to collect and publish in book form a series of papers and essays by his friend and ally W. H. R. Rivers, whose recent death had devastated him. The publication was partly to fulfill a promise to Rivers to make available to a general audience his memoirs on ethnological subjects that were scattered in journals. ES did this, publishing such landmark essays as his "Ethnological Analysis of Culture", "The Disappearance of Useful Arts" and "Sun-Cult and Megaliths in Oceania". But, as a more general tribute, he added in a number of Rivers's psychological and psycho-medical studies, including his renowned essays on "Freud's Concept of 'Censorship'" (which had made a big impact on ES) and "Sociology and Psychology", but also some unpublished notes and unfinished work (for which Will Perry collected bibliographical references). The collection was given the title *Psychology and Ethnology*.[10]

The essential theme of the essays was, in ES's view, the search for the psychological motives that impelled humans to develop customs and beliefs in certain ways: "For Dr Rivers, who was a psychologist first and an ethnologist afterwards, every problem in ethnology was essentially psychological" (preface, p. viii). What impressed ES was not only Rivers's groundbreaking insistence on exact scientific methods of data collection in ethnology, but his ability to cross boundaries. He leapt creatively from disciplines such as neurology and psychiatry to anthropology and sociology. ES focused Rivers's ethnology upon the theme of diffusion (for which he earned criticism of bias). At the same time he conceded that from about 1915 "Dr River was beginning to realize that he had pushed the theory of mixture of cultures further than the evidence warranted" (introduction, p. xvi). Thus he included some late essays (for example, upon circumcision, massage and sex in the Pacific) where Rivers gave modified and more comprehensive interpretations than formerly, for instance concentrating more upon magico-religious purposes and symbolism. However Rivers and ES remained sceptical of any theory (such as Freudian or Jungian theory) that postulated some innately derived universal system of symbolization among humankind. Rivers tested this hypothesis. He showed, for example, that some topics that were supposedly universal, such as rebirth as religious symbol or animism, were in fact not universal. The school of independent invention of culture commonly explained the existence of very similar customs in different cultures as due to the working of deeply embedded human propensities. Customs were explained as almost instinct-based responses to environmental challenges. Diffusion was almost automatically resisted.

If however, ES said: "we find that the use of a symbol is not universal, and still more if we find that the peoples who use it show other clear evidence of influence from without, the fact of diffusion becomes of great significance, and the probability greatly in favour of the view that, wherever it is found, it is part of the social heritage" (p. xxv). Rivers had thus paved the way towards eliminating much of "the darkness and the learned nonsense" from ethnology (p. xxviii).

Understandably, given River's tragic early death, reviews of *Psychology and Ethnology* were respectful. But there was scholarly scepticism about his diffusionism. The experienced anthropologist Brenda Seligman, for example, observed that the diffusion versus evolution controversy had tended to obscure some real issues. She agreed that human institutions must be studied in an historical context: "But however fascinating the study of history may be, and illuminating in elucidating the complexity of customs, the description of functions still remains the work of the ethnologist. The theory of diffusion leaves us with our main problem of human institutions unsolved". She pointed out the paradox that the Egyptians were supposed to have spontaneously developed "one culture in one place by one people from the simplest forms to a highly organized political entity", while this possibility was denied to other cultures in other places: "the world-wide spread of this [Egyptian] culture and its reception by savages, apparently persisting throughout the world in mental cold storage, is a hypothesis that seems to ignore all the psychological and economic needs of mankind".[11]

The 1920s were active years in archaeology. ES believed that a spate of discoveries from western Asia and India, Siberia, the Far East and America corroborated claims that he had been making since at least 1916. What had seemed to some people back then as hazardous guesswork now had more solid backing. He was particularly heartened by the work of Sir John Marshall, the discoverer of the Indus Valley civilization. Director of the Archaeological Survey of India (begun in 1861), Marshall had stunned the world with his excavations in the Ganges valley. His findings suggested close contacts between the ancient civilizations of Mesopotamia and the Indus Valley (publicized in 1926). As ES proudly pointed out in a letter to the *Times* he had plotted out the routes of cultural diffusion from the Euphrates–Tigris area to the Indus Valley a decade earlier.[12]

Another confirmation of his theories seemed to come from recently discovered Kaingaroan rock carvings in New Zealand. They depicted large canoes, which ES believed had significant affinities with vessels used in Egypt in the time of Tutankhamen. Depictions suggested the use of similar vessels in the Mediterranean from the thirteenth century BC and Sweden from about 1000 BC, prompting the claim that "the earliest maritime expeditions in Polynesia were probably made (in the early centuries of the Christian era) in ships of this type".[13] Raymond Firth, an expert on the Pacific (and one of Malinowski's students) immediately stepped in. Ever the diplomat, Firth praised ES but rejected his claim. There was, said Firth, no real identity of structural features between what was really only a variant of the ordinary Maori war canoe and ancient vessels. He concluded: "The wide scope of Professor Elliot Smith's researches into primitive shipbuilding is well recognized, but to adduce the recent Kaingarao carvings as a new 'confirmation' of his inferences will tend, in the eyes of New Zealand ethnographers at all events, to weaken the validity of his brilliant and suggestive theories".[14] In his combative way ES got enmeshed in disputes with experts over other discoveries, for instance, the finding of supposedly Aegean artifacts in Colchis in Georgia or Egyptian-type dual government systems in Bhutan in India or Shogunate Japan.[15] John L. Myres continued to attack ES's methodology (especially his fallible "method of agreement"). ES hardly improved their relationship by commenting: "It is surely inappropriate for a historian like Professor Myres, whose excursions into biological and physical science have often excited the wonder of his admirers, to complain because I do not dwell in a watertight compartment. His account of the evolution of the Mongolian race has made the Cambridge 'Ancient History' the most humorous book issued in the last decade".[16]

ES became exultant in the new year, 1927, when news arrived in London that J. Eric Thompson had located in the Ayer Collection of the Newberry Library in Chicago hitherto unpublished water-colour sketches made around the 1830s by the French artist Frédéric de Waldeck of designs on a sculptured wall of the Palenque Palace in Mexico. Palenque was a Mayan centre which reached its cultural peak around 500–700 AD. Waldeck had done engravings of Palenque in the early 1820s. He returned in 1832 and lived for two years on the ruined site, making over 200 drawings in water and oil colours, a project partly financed by the Mexican government. He later did more work in Yucatan, much of which was confiscated by the Mexican government after Waldeck had fallen out with them. Luckily the Palenque sketches were spirited away to France, where they languished for over thirty years. In 1860 a French government commission reported upon the collection, declaring the drawings far superior to others in existence, although they added that this was despite Waldeck having a "penchant for restoration" (i.e., he did not hesitate imaginatively to recreate some features so that audiences would have a better idea of the original; this was a common practice at the time, one of the best examples of course being Arthur Evans's "restoration" of the Minoan palace at Knossos in Crete). Waldeck's unpublished manuscripts and drawings were purchased by

Edward A. Ayer, a Chicago benefactor, who subsequently presented them to the Newberry Library. Eric Thompson, field director of the British Museum expedition in British Honduras, stumbled across the documents while engaged on research work at the library and "realized their importance as a possible source for the reconstruction of Maya history".[17] He generously permitted ES to have access to photographs of the sketches. ES arranged for the photographs to be splashed across the pages of the *Times* and the *Illustrated London News*, and he wrote feature articles about them. This discovery, he was convinced, "should settle once and for all the century-old controversy" regarding the identification of elephant-like creatures depicted in Mayan and Aztec ornaments and sculptures.[18]

Waldeck claimed to have discovered four bas-reliefs in a subterranean room in the Palenque Palace, from which he made water-colour sketches. (Unfortunately the originals disappeared during the following ninety years.) One of these, a design on a sculptured wall, showed what plainly seemed to be an elephant's head, in front view with mouth opened in the "bun-catching" attitude, set upon a serpent's body: "The lozenge-shaped form of the open mouth, the cut stumps of the tusks, and the markings on the under-surface of the trunk are all quite distinctive of the elephant", said ES in the *Times*. In an heraldic grouping the elephant figure was flanked by a conventionalized macaw and a tapir on a bird's body. The macaw and tapir were so different to the central figure that this seemed to dispose of those critics who insisted on substituting one for the other. Another photograph showed stucco slabs from Palenque, one of which pictured two elephant's heads in a floral shamrock-like design suggestive of the Chinese T'ang period. For this connection ES was indebted to his colleague John Collie (a chemistry professor at University College who was a connoisseur of antique Chinese and Japanese art): " . . . he has called my attention to the fact that the very un-Maya-like floral pattern intertwining both the elephants' heads" was like T'ang motifs of the eighth or ninth century AD, "a date that seems to be appropriate for the Maya building in which the slabs were found". This reinforced ES's theory of a cultural sweep from India and China to Central America. He drew a spectacular map for the *Illustrated London News* which had bold arrows indicating the routes by which Indo-Chinese culture reached Central America in the eighth century. Arrows went from India to Indo-China and Java, merging with arrows from China, then heading east, propelled by the northern equatorial current, passing through the Caroline group of islands across to Central America. Side arrows shot out to New Zealand, Hawaii and Easter Island.

To reinforce his case ES reproduced a photograph, taken by the Cambridge botanist W. Balfour Gourlay in 1920, of a crudely modeled stone representation of an elephant about two feet high that sat on a pedestal in front of the San Salvador museum. He also included reproductions of cylindrical vases found by Thomas Gann in 1916 in a mound at Yallock in Guatemala (which had been taken back to Britain in the fashion of the day). The painted pottery vases seemed to show two unmistakable elephants painted in natural colours: "the form of the head, body and legs does not admit of any doubt as to their identity as elephants; and the peculiarities of the lower jaw and teeth can be explained by studying the mode of conventionalizing elephants in Java and elsewhere in the Asiatic area" (*Times*). ES was now absolutely convinced that the Mayan artists were in fact depicting elephants and that the scenes portrayed represented the exploits of the Vedic god Indra, thus establishing Asiatic inspiration of early civilization in the New World: "The definite settlement of the elephant controversy marks a revolution in ethnology", he concluded his *Times* article. Ancient Indian art had reached its zenith from the fourth to the ninth centuries AD. Gupta art (circa 350–650) had exerted pervasive influence in Indo-China. The elephant-headed god Ganesh became almost universal in south-east Asia, and India left its mark on Chinese Buddhist art generally (as also during the great Nara period in Japan). From there the great cultural current flowed across Oceania to America.

ES's "decisive victory" turned out to be less than Napoleonic. This was invariably his experience regarding America. H. J. Braunholtz, assistant keeper of the ethnological collections at the British Museum and an important figure in the Royal Anthropological Institute, questioned Waldeck's reliability as a draughtsman. This, he said, could be tested by comparing Waldeck's numerous drawings of ancient Mexican monuments (published in 1866, but not including his sketches of the Palenque slabs) with photographs of the same subjects in A. P. Maudslay's respected *Biologia Central-Americana: Archaeology* (volume 4, 1896–1899).

> Such a test demonstrates beyond all question that Waldeck, although comparatively faithful to the originals in a general way (and perhaps unusually so for the period at which he worked), was not only grossly inaccurate in many important details, but often introduced considerable modifications of an imaginary kind. In particular he displayed a tendency to transmute the well-known Maya serpent motive into something of an elephantine character.

Braunholtz warned against accepting Waldeck without independent corroboration.[19] Writing from Luxor in Egypt, Maudslay also spoke of Waldecks' "inaccuracy and the worthlessness of his drawings" in support of ES's views.[20]

ES replied robustly to Braunholtz's criticisms. He pointed out that fifty years had elapsed between Waldeck's pioneering work and Maudslay's much more complete investigation, during which time the Palenque artifacts had suffered badly. More to the point Maudslay had not had access to the objects now under discussion. He cited internal evidence of the essential accuracy of Waldeck's sketches in this case. Even if it were granted that Waldeck had "imagined" his elephants, how could one explain the T'ang-like floral patterns: "If Waldeck invented the designs that are under discussion, how are we to explain the precision with which he reproduced the patterns that were fashionable in Cambodia and Java at the time Palenque was being built in America, when neither the designs themselves, nor the fact that they were contemporaneous with the Maya reliefs could have been known?". Again how could Waldeck have invented decorative details on the elephant's head (headband, a rosette on the forehead) that were identical with those found on Javanese sculptures of the ninth century? Waldeck knew nothing of these matters in the 1830s. Finally, ES pointed out that "Mr. Braunholtz and his colleagues in the British Museum" had refrained from mentioning the San Salvador sculpture or the vases from Yallock.[21]

Subsequent research has indicated that "Count" Waldeck's credentials were indeed problematical. He was a flamboyant and mysterious figure. Some have seen him as a great scholar, others as a charlatan. He was probably a bit of both. He claimed to have been born in 1766 (which would have made him 109 when he died in 1875), to have studied under the famous Napoleonic artist David, and to have traveled to Egypt with Napoleon's expedition, none of which can be documented. He clashed with the Papacy for publishing an ancient set of pornographic prints and had a turbulent, if adventurous, time in Mexico. In 1825 an English mining company in Mexico hired him as a hydraulic engineer. His interests turned to the Mayan sites of Palenque and Uxmal. He was commissioned to do engravings of the sites by Lord Kingsborough, a man obsessed by ancient American antiquities. Kingsborough had fantastical theories about Mesoamerica, including the old chestnut that Native Americans there were descendants of the Lost Tribes of Israel.[22] Waldeck was accused, even at the time, of making the Palenque ruins look more Egyptian than they were in order to please Kingsborough. In Paris in 1838 Waldeck published a book of illustrations of Mayan ruins, including an Uxmal pyramid, again seemingly in the Egyptian tradition.

In a later publication of 1866 his illustrations of panels of Maya script in the Temple of Inscriptions at Palenque (also mentioned by Braunholtz) "included clear depictions of heads of elephants (now known to be erroneous embellishments). This fueled speculation about contact between the ancient Maya and Asia and the role of the mythical lost continent of Atlantis as a common link between ancient civilizations of the Old and

New Worlds".[23] ES carefully avoided discussion of "the mythical lost continent of Atlantis", which he clearly thought to be nonsense. Whatever the truth about Waldeck's "embellishments", and however dubious his life story, questions raised by ES about the Palenque sketches unearthed in 1927 remain to be answered.

Eric Thompson gave somewhat ambivalent support to ES when he published a lucid account of his discovery of the "elephant heads" in *The Scientific Monthly* in November 1927.[24] Thompson is another fascinating figure. He was to become a dominant, even domineering, figure in Mayan studies. It is the common opinion now that he was the main obstacle to the deciphering of the ancient Mayan scripts, not achieved until after his death in 1975. (The story is brilliantly told in Michael Coe's *Breaking the Maya Code*, 1992.) Back in 1927 Thompson accepted that Waldeck held preconceived ideas on the Asiatic origin of Maya civilization, and that some of his drawings were discrepant from Maudslay's more accurate photographs and drawings: "However, the boldness of the outline of the first strip [the full-face elephant] would seem to point to the fact that he was copying from a clean and undeteriorated original" (p. 393). Overall (if Waldeck was reliable) here were "three perfectly good elephants' heads" (p. 392). Thompson outlined the debate between the diffusionists and the "hands off America" school dominant in the US. He came down rather in the middle: "Even should it be admitted that elephants' heads were portrayed on these Maya carvings, the diffusionist argument would be as far from proved as ever. The attitude that American civilization was never affected to any appreciable extent by Asiatic influences would have to be abandoned, but there is a great difference between conceding this point and admitting that the whole of American civilization was of Asiatic origin" (p. 395). The facts of agriculture and metallurgy did not support the "Children of the Sun" theory. His view was that long after New World civilization had gotten under way, influences that crossed the Pacific profoundly affected it, especially in the realms of religion and art. He called for less verbal abuse between the competing schools and for more systematic excavation. No doubt many scholars felt the same.

ES may well have been tiring of the American quagmire by now. So he raised his sights. He gave an address on "The Philosophical Background of Ethnological Theory" for the *Journal of Philosophical Studies*, and one on rationalism for the *Rationalist Annual*. In the first paper he accused modern ethnologists of being essentially unscientific. They were in the tradition of Cartesian scholasticism; and that was at core isolated from the whole principle of modern science.[25] It was not a particularly convincing argument and professional philosophers wisely ignored it. His talk to the Rationalist Society explained the neurological basis of rational thought and gave evolutionary reasons for the development of the greater intellectual powers of humans compared with their ape relatives. ES sketched the origins of religious practices in early human cultures. He concluded: "Religion is essentially a sublimation of the life-struggle".[26] In one of his last writings he would declare: "The satisfaction of the craving for life and resurrection is the aim of every religion. The fundamental instinct of self-preservation does not only find articulate expression in religion, but provides the motive of all mythology and folklore. Religion, mythology, folklore are all so many varied expressions of Man's craving for safety and prosperity".[27]

The year 1927 also saw the appearance of two of his more general pamphlets: *Human Nature* and *The Beginning of Things*. The first was the Conway Memorial Lecture, with Charles Darwin's son Leonard in the chair. Leonard Darwin was a leading light in the Eugenics Society, and like many eugenists, inclined to ideas of genetic determinism. ES had inclined towards eugenics at times, but always believed "that both nature and nurture play a part in moulding personality"; that the educational and social environment was at least as important as heredity and innate differences. He stated openly to his South Place audience that he generally endorsed J. B. Bury's idea – expressed in his popular book *The Idea of Progress* (1920) – "that the social inheritance of ideas and emotions to which the individual is submitted from infancy is more important than the tendencies physically transmitted from parent to child".[28] Most of *Human Nature* was devoted to the proposition "that man is by nature a kindly and considerate creature, with an instinctive tendency to monogamy and the formation of a happy family group bound together by mutual affection and consideration" (p. 41). ES repeated the argument made in *The Evolution of Man* that early humans and most primitive peoples were essentially peaceful. He cited Perry's ethnological research, listing an impressive host of primitive cultures that were supposedly unwarlike. They ranged from the Punans of Borneo, the Negritos of the Phillipines, the Aru islanders, the Kubu of Sumatra, Andaman Islanders, Veddahs of Ceylon to jungle tribes of southern India, the Pygmies and Bushmen of Africa, the extinct Tasmanians, many Australian tribes, many American Indian groups (such as the Dene, Salish, Northern Ojibway, Paiute) and Eskimo peoples. Such examples were to find their way into the booming peace literature of the twenties and thirties and into some anthropological literature. The propositions of a Golden Age of peace, the essential peacefulness of indigenous cultures, the claim that war was avoidable and could not be attributed to "human nature" – these were as vigorously disputed by critics.

The year closed with the doyen of Oxford anthropology Robert Ranulph Marett making a characteristically elegant but bitter attack upon ES.[29] Marett, although himself a Jerseyman who for long felt himself an outsider in Oxford, had succeeded E. B. Tylor as Reader in Anthropology at Oxford in 1910 (anthropology was deemed not worthy of a chair) and by dint of hard lobbying headed the new Department of Social Anthropology set up in 1914.[30] He was an expert in "primitive religion" or religious ethnology. He made his name revising Tylor's influential theory of "animism" as the prime factor in early religion, but retained enormous respect for Tylor and that other icon in the field James Frazer, author of the hugely popular book *The Golden Bough*. Marett was incensed that ES consistently ridiculed the theories of Tylor and Frazer and, in the appropriate forum of the Frazer lecture, defended them.

ES had aggressively declared Tylor's doctrine of animism to be "the merest guess" at a plausible explanation in flagrant defiance of the known facts, and Frazer's work to be that of a disciple of Tylor highlighting the worst of his fallacies. Marett felt rather like a Greek champion defying the Persian barbarians at Thermopylae. But he screwed up his courage and defied the "iconoclast" ES. Surely, he asked, "no one who has studied *Primitive Culture*, with its vast array of evidence so lucidly digested, will venture to assert that Tylor scamped the preliminary induction on which his theory rested".[31] There could be no justification for holding Tylor up to scorn "as one who ignored evidence that was not yet to hand. If Prof. Elliot Smith is really guilty of this anachronistic attitude to Tylor, then as a critic he is past praying for" (p. 176). On the more general issue of diffusion of culture, Marett was sceptical of extremes. Of course, he admitted, much could be learnt from the spread of cultures, but it was unlikely that diffusion of culture afforded a full explanation of the nature of culture (a view he attributed to the Smith–Perry school): "I want to put the argument from diffusion in its place as part of a more comprehensive and complex apparatus of research" (p. 173). The Children of the Sun saga was fine as a working hypothesis, if put forward with proper reservations. Interestingly Marett was sympathetic to Perry, who "puts his case moderately and with a growing sense of the difficulties that he has to meet", but "I cannot see how the premature dogmatism of Prof. Elliot Smith is going to be of the slightest service either to the particular cause which he has made his own or to the cause of science in general" (p. 179). Marett's general portrait of ES was of a man intoxicated by "the fumes of a world-embracing fancy", a man who prejudiced his cause "by allowing exaggeration and bias to pervade his forensic efforts" (p. 181), one who crammed his ideas down peoples' throats. Such slighting language was rare for an academic lecture. Marett finished on a placatory note: "I would add that in real life Prof. Elliot Smith has always struck me as a most reasonable as well as otherwise charming man. It is only on paper that he appears to me to let his enthusiasm run away with him" (p. 189).

Not long afterwards Roland B. Dixon made a wide-ranging attack upon ES's American theories in *The Building of Cultures.* Dixon, a respected figure, worked at the Peabody Museum of Archaeology and Ethnology at Harvard, which was a hotbed of opposition to any idea of external influences upon indigenous American cultures.[32]

CHAPTER

6

The Diffusion Controversy (1928–1933)

Malinowski felt in 1926 that diffusionism was "now fashionable". A number of anthropologists, he observed, not without irony, "are busy reconstructing the influence of Egyptian culture on Central America; they quarrel as to whether all civilization started in Mesopotamia, Atlantis, or Pamir. This historical or diffusionist trend is now being advertised as the 'revolutionary' or 'modern' school of anthropology, though in reality it is as old as the Ten Lost Tribes fallacy".[1] Certainly ES was still supremely confident of the soundness of his position. He and Malinowski slugged it out briefly in the pages of *Forum* in a debate labeled "Is Civilization Contagious".[2] Both reproduced their papers in a book published in 1927 entitled *Culture: The Diffusion Controversy*. The book was strangely unbalanced and not in the diffusionists' favour. There were short essays by ES, Malinowski and Alex Goldenweiser, together with a fifty-one page monograph by Herbert J. Spinden.[3] Three to one against ES. It was perhaps a sign of the times.

ES reiterated his now familiar case for diffusion. Why should not cultural traits have been spread in ancient times just as they had been in modern times, as in the case of inventions such as typewriters? It defied belief that independent invention could explain marked congruences between cultures: "For if any community can of its own initiative create a civilization, a more difficult problem has to be solved: why it acquires a multitude of features in its arts and crafts, customs, and beliefs that present a striking similarity to those of other communities, when all considerations of contact or prompting directly or indirectly are excluded" (pp. 9–10). Egypt was a prime candidate as the original fountainhead of later civilizations. It was admitted even by opponents that Egypt had exerted great influence upon European civilization, via Crete, Asia Minor, Greece and Rome. Recent archaeological research had shown growing evidence of Egyptian/Mesopotamian influence upon

India. This was exported (ES believed) to South East Asia and the Pacific. There was also recent evidence of the diffusion of culture from Sumer and Elam to places like Turkestan (Pumpelly's dig at Anau had indicated this) "right up into the heart of Siberia and into the Shensi Province in China . . . these people in the Far East were making arrow-heads of chalcedony and other flint-like stones: also other stone implements, rings of stone and shell, beads, pottery (both monochrome and painted), and even small figurines, all revealing clear and unmistakable indications of diffusion of culture from Mesopotamia" (pp. 18–19).

As for America, it was "an altogether incredible supposition that the Polynesian sailors who searched many thousands of miles in the Pacific with such thoroughness as not to miss even the minutest islets were not repeatedly landing on the shores of America for ten centuries or more. How could such people who found Hawaii, Easter Island, and New Zealand have failed to discover the vast continent stretching from pole to pole?" External influence could explain the sudden appearance of civilization in Central America, Mexico and Peru. And this civilization "conformed in almost every respect to the distinctive type of civilization (admittedly a very peculiar one) that was flourishing in the southeastern corner of Asia at the time when it made its appearance in Central America. The type of pyramid found in America was also the dominant feature of the architecture of Cambodia and Java during the same centuries. The same system of beliefs and customs, the same distinctive features of its architecture, in fact a whole series of arts and crafts, customs and beliefs, were suddenly introduced into the New World, which seem to bear unmistakable evidence of their Asiatic origin" (pp. 23–24).

Malinowski, the up-and-coming *guru* of social anthropology, sounded virtually magisterial in his rebuttal of ES. Of course diffusion of culture occurred between communities. Nobody denied it. But to suggest that the essence of cultures derived from some single original source was ridiculous. The very concept of a sharp distinction between diffusion and invention was in reality misleading. Even modern "inventions" had been made and remade in different times and places, witness the debates over who really invented the calculus, the steam engine, the wireless, etc. It was especially complex with cultures: "Just because no idea and no object can exist in isolation from its cultural context, it is impossible to sever mechanically an item from one culture and place it in another. The process is always one of adaptation in which the receiving culture has to re-evolve the idea, custom, or institution which it adopts" (p. 31). Malinowski stressed the versatility and ingenuity of indigenous people when it came to solving problems or meeting needs: "The existence in social organization, in religion, in language, and in economics of cultural contrivances which satisfy the same need, which are thus functionally akin, and which bear an entirely different physiognomy and are carried out by entirely different mechanisms, spells all over the surface of human culture the assertion of independent origins" (pp. 33–34). Sex and economics were the real keys to understanding the life of culture. Malinowski explained how this could be discerned in such things as courtship, mating, family organization, even religious and artistic values. He concluded that of course diffusion must be studied, trade routes examined, etc. But diffusion never took the form of mechanical transmission. It was always a readaptation, "a truly creative process, in which external influence is remoulded by inventive genius". No culture was a simple copy of another: "The culture of Egypt is no older than that of China, Mesopotamia, or India, and it took as much from its neighbours as it gave" (p. 46).

Goldenweiser took a theoretical stance generally similar to Malinowski, although he did dissent from the "inexorable determinism" that underlay Malinowski's view that all culture was driving humanity upon a path of inevitable progress (p. 104).

Spinden distinguished between what he called the "prosaic" versus the "romantic" school in anthropology. The prosaic school believed that there were "common psychic trends if not complete psychic unity among all the races of mankind. Similar experiences are everywhere producing similar results in handicraft and statecraft." This school accepted "the possibility of independent invention – of thoughts finding expression over and over again – and sees no need for straining historical evidence beyond the elastic limit to account for the dissemination of certain cultural traits around the world". The romantic school, on the other hand, "sees in cultural similarities the almost certain proof of dissemination from a favoured intellectual source even though ages and oceans intervene". The romantics, who were in reality a diverse and divided lot, pictured the great mass of humanity as devoid of inventive ability but possessed of an extremely retentive memory: "Ancient transmissions and inoculations, of which history furnishes not the slightest direct evidence, are invoked as a logical necessity where there is any detail to be exploited as a surviving strain" (pp. 48–49).

The romantics tended to be gullible, belonging to the fantasist tradition that suggested (for example) that the Toltec god/king Quetzalcoatl (*c.*1200 AD) was St Thomas, that the Ten Lost Tribes of Israel had settled in America, or believed in the lost continent of Atlantis. A number of nineteenth-century writers, followed by the German diffusionists (such as Graebner and Schmidt) had preceded ES, Perry and the worthy but deluded Rivers in suggesting an Egyptian or Asiatic source for Central American civilization. Spinden mocked ES's speculations about elephants in American art. He was reviving old identifications, inventing new ones and relating these "supposed representations to Buddhist pictures, also dilating on certain grotesque, composite figures of southern Asia which he holds to be the first parents of all American monsters". Such theories, "sincere and hard-working enough, are based on the narrow and depressing concept that man in general cannot think for himself but must imitate and remember the actions of a favoured race" (pp. 59–60).The rest of Spinden's article was given over to a detailed account of how independent civilizations arose in America. He dealt with prehistory, comparisons of New and Old World agriculture (maize versus wheat based), ceramic and textile arts, metallurgy, the calendar and medicine (not a single important disease of parasite type was common to the New and Old Worlds until Columbus). Nowhere did he find significant evidence of Old World influence upon American civilization before the Spaniards.

Herbert J. Spinden was an opponent to be reckoned with. He was a Mayan expert, whose *Study of Maya Art* (1913) became a classic. Harvard-trained in archaeology, he was curator of that university's Peabody Museum from 1921–1929. He spent much of his life perambulating the jungles of Central America (including a long stint in the early 1920s in Guatemala, Honduras and Nicaragua), carefully examining buildings, artwork and other artifacts. A corpulent man with shaggy white hair, he was a colourful figure and considerable raconteur. He was to become an almost avuncular figure in American archaeology. Personally generous and modest, he could be formidable in print. He needed to be, as his own theories on the Mayan calendar were controversial, and have (like many other theories of the time) been largely superseded by recent science. His views on early agriculture and the origins of Mayan civilization have turned out to be more durable.[4]

ES continued to work on his Grand Design, an holistic interdisciplinary centre that would combine the biological and social sciences with the humanities. Its hub would be London (at University College obviously) but with tentacles reaching out to British Empire countries such as Canada, Australia and South Africa. The aim would be to break down barriers between disciplines such as anatomy, physical and cultural anthropology and psychology, to encourage innovative thinking and research, all under the umbrella of a term such as "Human Biology". As was already happening at University College, medical students should be taught not only anatomy and neurology but also anthropology and psychology, while anthropology and psychology students should learn about brain neurology and the central nervous system as well as the study of customs and beliefs. Cultural anthropology should be an integral part of psychology and vice versa. ES envisaged schools, or institutes, of anatomy, bio-medicine, psychology and psychiatry, constantly interacting and sharing information and concepts, a foreshadowing of today's bio-medical technologies. (Like some of today's advocates of "human enhancement", ES also felt he was constantly being frustrated by bio-conservatives with outdated outlooks.)

In a letter to a colleague which gave a detailed account of his futurist dream he outlined a sort of super-university:

> An Institute in which there would be laboratories for comparative neurology, for experimental psychology, for physical anthropology in the ordinary sense of the word, and for geography, linked up with a hospital that would deal with neurology, ophthalmology and psychiatric cases, would represent the germ of a post-graduate University, whose centre of interest would be definitely the study of Man, his history, behaviour and beliefs. It would eventually involve the study of the humanities in a much more rational and fruitful way than is done at present. For cultural anthropology would necessarily come into contact with and play a part in the interpretation of classical archaeology and mythology, as well as of history and literature, and in the interpretation of symbolism in general.[5]

Unfortunately funding never became available for this ambitious project. The 1929 world Depression extinguished any lingering hopes he may have had for it. The Depression and the rise of the Nazis, with their pseudo-scientific Aryan race theory, made the thirties a grim time for ES (as for other liberal thinkers). His failing health from 1932 only made matters worse. But he struggled on against what he took to be the stupidities and evils of the time.

It is likely that his ultimate health problems stemmed from overwork in the mid to late 1920s. As usual he was hectically busy. A simple listing will indicate this. Apart from his heavy schedule at University College, he gave a series of guest lectures, including a course at the École de Médecine at Paris in 1925; another set of lectures on early man and the growth of civilization in 1927 at Gresham College, London, that was to become his bestselling book *Human History* (1930); the Conway Memorial Lecture in 1927, the Huxley Memorial Lecture in 1928, and numerous other addresses. He immersed himself in palaeoanthropological issues, studying and analyzing the hominid fossils that had been unearthed with the discovery of *Australopithecus* in Africa, possible ancestors of humans. He attended the Pacific Congress of 1929, held in Java.

There he joined in discussions of "Java Man", bones of an ape-like man famously discovered by the Dutchman Eugène Dubois in 1891 at Trinil in Java that he named *Pithecanthropus erectus*, now more commonly classified as *Homo erectus*. (Three years later ES was to address the International Congress of Prehistoric and Protohistoric Sciences held in London, updating them on *Pithecanthropus*.) In 1930 the Rockefeller Trustees invited him to travel to China to examine the fossil remains of "Peking Man" or *Sinanthropus*, excavated in 1929 by Pei Wenzhong at a palaeolithic site in Hebei Province. Rockefeller were funding the excavations. He accepted eagerly. He lectured widely on his experiences and thoughts regarding this highly significant collection of *Homo erectus* fossils. He famously presented his findings, indicating a possible Asian origin for humans, at the Zoological Society in London in 1931. In doing so he unknowingly upstaged the less flamboyant Raymond Dart, who followed with a modest presentation on the "Taung skull" (*Australopithecus africanus*):

> At the meeting Elliot Smith launched into a masterful exposition of the significance of Peking Man, replete with projection slides and casts for the audience to handle as he dazzled them with evidence of Peking Man's upright stature, fire-making abilities, and propensity towards cannibalism. It was a tough act to follow.[6]

What is clear from such descriptions is that by this time ES had become what is now labeled a "public intellectual". He was constantly touring, giving public lectures, talks and press interviews, writing articles for newspapers and popular magazines as well as learned journals, even using the new-fangled medium of radio.[7] During the 1920s and 1930s he had become a guru on topics such as human prehistory, the way the human brain and nervous system worked, and of course Egypt and cultural diffusion. He was regularly trotted out to commentate when new discoveries were made or issues raised in such discourses. He never quite captured the stature of public intellectuals such as Bernard Shaw or H. G. Wells, who were hyperactive at the time and had immense exposure in what we now call the media. But ES was not far below them. If he lived today he would be a regular on television. He was the equivalent in science matters of today's spokesmen like Richard Dawkins or Paul Davies. This could be a plus and a minus, a plus in that he had a platform for his ideas, a minus because the academic world tended to be disdainful of such populist exhibitionism.

ES's book *Human History* came out in 1930. It was probably his most popular book. It was a blockbuster 509 pages long, written in his usual readable style, although there were large slabs of technical detail that general readers must have found hard going. It was a syncretic history of humanity. Such works were thin on the ground at the time, the most famous being H. G. Wells's *Outline of History* (1920), which began with the creation of the universe and took the story to the present. Both works gave considerable space to the evolution of life, prehistory and ancestors of man (ES's book devoted six of its fifteen chapters to the period before the rise of civilization, dated from about 4000 BC). If Wells's writing was more exciting on these subjects, ES's expertise was superior. However the reading required to update his information in these massive discourses was gargantuan, and he eventually decided to finish with the Greeks and their legacy, leaving medieval and modern history to others. The book was in essence a fusion of his *Evolution of Man* with his diffusionist writings, reinforcing his thesis wherever possible from recent discoveries, expanding here, condensing there (for example, he kept his commentary on America to a minimum). He drew on the ideas of an impressive list of archaeologists, ethnologists, classical scholars and historians who were writing in the 1920s. They included V. Gordon Childe (whose *Dawn of Civilization* had appeared in 1928), Leonard Wooley (whose excavations at Ur had created a sensation), Wilfred Le Gros Clark, the Seligmans, A. P. Elkin, Arthur Hocart, C. Darryl Forde, Baldwin Spencer, Franz Boas, T. H. Massingham, Donald A. Mackenzie, P. N. Ure, Solly Zuckerman, Gilbert Murray, F. M. Cornford, Charles Oman and H. A. L. Fisher. He relied on the Cambridge histories, especially *The Cambridge Ancient History* (1923), as reference works.

What ES was saying in this book, which in a sense summarized his lifework, was this: that diffusion was not a crackpot theory but an essential feature of life. The whole of humanoid and human history was the history of migrations of peoples and cultures. Peoples carried their cultures into new areas, where interaction constantly took place with existing cultures. Influence, adaptation and assimilation were the order of the day. As we have seen, neither ES nor Perry claimed that civilizations were transplanted holus bolus. Nor did they speak of the spread of mechanical elements or items but of the spread of customs or clusters of customs, behaviours, arts and techniques. These were taken up and used by the acceptor groups where they were of use, either for brute survival or for more complex reasons of ritual, religion, and so on, all of course ultimately of evolutionary use. Gradually new cultures arose that were an amalgam of old and new knowledge, and this had been happening from the time of hominids till Egyptian civilization and beyond. Diffusion was the dynamic reality of history, while the anti-diffusionist worldview of spontaneous invention and constant resistance to ideas of migration of cultures was in fact an unhistorical and basically static worldview.

Unfortunately the legacy of Descartes had made an unreal schism between the study of nature (science) and the study of mankind (the humanities). Modern anthropology still laboured under this false dichotomy. ES's book aimed to reunite the two discourses, to fulfill the remit of his University College school, to create a genuine discipline of "Human Biology", to combine biology with history to give a broader interpretation of human thought and action. The theories of geology and biology were founded upon the principle of continuity and this needed to be applied (or reapplied) to the study of mankind. Human history could only be interpreted fully if new emphasis was placed upon the principle of continuity of culture in both time and space, and this could only be expounded by the historical method.

The study of anatomy and physiology gave essential insights into human culture, yet historians and anthropologists were profoundly ignorant of such things. However as ES said:

> Several millennia before Man systematically studied anatomy he was building up a fabric of civilization under the influence of doctrines based upon his ideas of the functions of the heart and blood, the breath and moisture, the placenta and the hypothetical 'life-substance'. It would, in fact, not be an exaggeration to claim that civilization was evolved out of Man's endeavours to understand the constitution of his own body and to preserve the life that animated it. (p. 13)

Hence *Human History* wove an intricate story dealing with such topics as mummification, tombs, megaliths, the use of metals for weapons and tools, the search for "life-giving" precious stones and metals, the spread of a heliolithic culture complex, hammering home to a more general audience the themes he had been expounding in papers, lectures and books for thirty years.

Overall, although the book was essentially corroborative of his Egyptian theory, there was a discernible sense of growing disillusion with ancient Egypt. As we have seen ES was never an uncritical admirer of Egypt, although still proclaiming its global significance as a purveyor of culture. All along he saw the rise of civilization (identified with Egypt, but with possible claims for Sumeria and Elam) as a two-edged weapon for humanity. It ushered in material and artistic progress on a massive new scale, but it also spelt doom to the supposed "age of innocence" and peace that had preceded it. It ushered in the state system, complex social organization, imperialist and militaristic forces, warfare and class conflict.

These forces were still at work, as tragically indicated by the rise of expansionist nationalism, and racist authoritarianism from the later nineteenth century, and the catastrophic war of 1914–18. Nor had the war put an end to such evils, as shown by the divisions and ethnic hatreds in Europe and – something that ES was quick to sense – the rise of totalitarian movements such as Fascism and Communism. Despotic ancient Egypt was looking less and less like a desirable model. In his book ES concluded that the rise of Greece, especially Athenian civilization – individualistic, rational, sceptical and scientific – was a necessary step in emancipating humanity from the shackles of superstition and tyranny. As he said:

> When Man began to devise Civilization, he became entangled in the shackles of the theory of the State, which he himself had forged.
>
> It remained for the Greeks to remove the shackles and restore to human reason the freedom it had lost.
>
> Ever since then the history of the world has been a conflict between the rationalism of Hellas and the superstition of Egypt.
>
> It depends upon the human population of the world themselves to decide which will win. For thought and courage can decide the issue.[8]

The book was generally praised for its scholarly merits and popular style (one reviewer thought it "so far above Wells' *Outline of History* that there is no comparison"[9]) but it was criticized as too speculative and sweeping.[10]

Publications continued to roll off the assembly line. In 1931 ES contributed an essay on the evolution of man to a collection of lectures delivered to the Royal Anthropological Institute entitled *Early Man: His Origin, Development and Culture*.[11] In it he concentrated upon the anatomy of early human fossils such as *Sinanthropus*, and updated themes that he had dealt with in *The Evolution of Man*. He now placed even more emphasis on the evolution of stereoscopic vision as a precondition for the development of human consciousness: "Man's intellectual preeminence is based primarily on the evolution of macular vision in a Primate with adaptable hands, which attained the erect attitude when the cerebral cortex under the conscious influence of vision came to control and regulate posture" (p. 42). Overall he concluded that the spectacular fossil finds of the last eighty years afforded tangible evidence of the diversity of the human family and of the range of its early wanderings. His cultural diffusion theory flowed directly from these findings. It was not, as so many have assumed, some quirky and irrelevant side interest that ES had been bewitched into because of a youthful and romantic love affair with Egypt.

In the same year, 1931, he produced *The Search for Man's Ancestors*, aimed at a popular audience.[12] It again dealt with the finding of important fossils such as *Pithecanthropus* and *Sinanthropus*. Wilfrid Le Gros Clark, who was to become a great authority in the field, was impressed: "With happy combination of erudition and clear expression, Professor Elliot Smith has produced, within a shilling manual, one of the most important summaries of recent research in human palaeontology . . . the author gives us a critical study of the significance of the fossils in the light of recent anatomical research, and we welcome the mature judgment and methodical reasoning on matters which often lead to hasty generalizations and bitter controversy".[13] Soon after there followed an expanded and illustrated edition of his small book *In the Beginning*, in the popular series *The Thinker's Library*. The main additions were on the topics of totemism and exogamy in ancient Egypt, with new reflections on "primitive religion", especially on E. B. Tylor's theory of animism.[14] ES had been thinking anew on Tylor and was soon to publish more on him.

During the summer of 1932 he wrote what was to be his final work on diffusion, *The Diffusion of Culture*. He saw it as a largely historiographical study of some major opponents of diffusion, mainly E. B. Tylor, William Robertson and W. H. Prescott. He had long reflected on Tylor's "malign" influence, most recently in his little book *In the Beginning* and in a short essay for the compilation *The Great Victorians* (1932). He wrote to his friend Donald A. Mackenzie on 26 September 1932:

> I have just sent off the MS. of a book. It is essentially historical – a line of approach for which I am wholly indebted to you, when one day in a second-hand shop in Edinburgh, you bought me Robertson's *History of America*.
>
> The greater part of the book consists of an analysis of the writings of Robertson, Prescott and Tylor to discover how each of them "went off the rails".

The great paradox, as ES saw it, for someone like him, wanting to concentrate on the dynamics and actual evidence of diffusion of culture – something in fact that most people would recognize as obvious because it was happening to them every day of their lives and was a conspicuous feature of history – was that he must first sweep away the ethnological misconceptions, the mythical will-o'-the-wisps, that blocked the path of rational inquiry. As Tylor himself had put it (ES liked quoting Tylor against himself): "a once-established opinion, however delusive, can hold its own from age to age".[15]

ES had little trouble disposing of the "delusive opinions" of the eighteenth-century scholar William Robertson, principal of Edinburgh

University. Robertson's widely read *History of America* (1777) examined the origins of American civilizations and contemptuously dismissed any idea that their customs may have derived from an external source. "Wild and chimerical" theories of course abounded, including the Atlantis fantasy. Some imagined that the inhabitants of the American continent were a separate race of humankind, or were antediluvian survivors of the Biblical Deluge. As Robertson said:

> There is hardly any nation from the North to the South Pole, to which some antiquary, in the extravagance of conjecture, has not ascribed the honour of peopling America. The Jews, the Canaanites, the Phoenicians, the Carthaginians, the Greeks, the Scythians in ancient times, are supposed to have settled in this Western world. The Chinese, the Swedes, the Norwegians, the Welsh, the Spaniards, are said to have sent colonies thither in later ages, at different periods, and on various occasions. (quoted p. 52)

He regarded as "frivolous or uncertain" attempts to show that customs travelled from other continents to America by tracing resemblances between the manners and behaviours of the societies being compared.[16] What Robertson was saying, essentially, was that in cases of resemblance the presumption should not be diffusion of culture, but rather that of natural response to similar environments: "If we suppose two bodies of men, though in the most remote regions of the globe, to be placed in a state of society similar in its degree of improvement, they must feel the same wants, and exert the same endeavours to supply them. The same objects will allure, the same passions will animate them, and the same ideas and sentiments will arise in their minds". Robertson was an early proponent of the "psychic unity" theory of humankind and also the "stage theory" of evolution. Humans were essentially similar, and they had inevitably gone through the same stages of evolutionary progress: "In every part of the earth the progress of man hath been nearly the same, and we can trace him in his career from the rude simplicity of savage life, until he attains the industry, the arts, and the elegance of polished society. There is nothing wonderful, then, in the similitude between the Americans and the barbarous nations of our continent" (quoted pp. 53–54).

Robertson accepted that not *all* customs could be explained as natural responses to environment. There were some that seemed "usages of arbitrary institutions", where there seemed almost accidental resemblances with other cultures. However he dismissed such cases as "few and equivocal". It was ES's whole thesis that this was not factual, and he ransacked his earlier writings to demolish Robertson's overall argument. (As a book like *Elephants and Ethnologists* was out of print, he felt justified in repeating large slabs of it verbatim.) He rehearsed his familiar arguments and examples, but added new ones. Robertson had asked why Americans had not embraced highly useful European inventions, such as the use of iron or the plough, if there had been cultural contact. ES replied that it was well-known historically that a people borrowing culture from another society did not necessarily adopt all its arts and crafts:

> To cite one instance, the Polynesians, who, as every ethnologist admits, derived their culture from Asia at a time when many of the so-called 'necessary arts' were in common use there, failed to adopt many of them. As they represent the chief agents concerned in the transmission of Asiatic culture to America it is important not to forget the lessons of Oceania. In particular the Polynesians did not adopt wheat and barley or the wheel, the absence of which in pre-Columbian America is constantly cited as an insuperable difficulty in explaining the origin of American civilization by diffusion. (p. 59)

He cited yet again Rivers's contention that many useful arts had in fact been lost over time after diffusion had taken place (something Robertson had thought impossible). Another problem with Robertson was that he confused races with culture, ignoring the fact that numerous carriers of culture, different peoples of different "races", were the agents of diffusion. ES gave a graphic example, that of Roman-style armour. In the text ES included a sketch he had made of a wooden figure from Hawaii wearing a distinctively Roman-type helmet (p. 65). German and other ethnologists had long drawn attention to the existence of Roman-style plate armour in Japan, China, the Pacific islands and north America. These could hardly have been the result of the "spontaneous combustion" of independent origin (he borrowed the "spontaneous combustion" parallel from Ratzel, who thought that the idea of independent origin was the anthropological equivalent, not of evolution, but of the discarded biological concept of spontaneous generation). Roman soldiers had certainly got as far as India. Some had acted as bodyguards to Tamil kings. More importantly colonies of Roman subjects had traded and settled in southern India. Whether taken directly by people or indirectly by commerce, Roman (or more generally Hellenic) products (such as armour) or ideas became established in India, and were then spread abroad widely as the result of Indian cultural expansion and influence. It was a familiar story "how the influence of the Indian Gandara and Gupta phases of culture, saturated as they were with Graeco-Roman traits, spread from India to Indonesia, China, Japan, the Malay Archipelago, and Oceania, particularly during the eighth and ninth centuries A.D." (p. 68).

Transmission to the American continent could have taken place either by an essentially land route via Turkestan and Siberia, or by sea routes across the Pacific. He piled on even more recent evidence of the marathon voyages and sea-faring capacities of Pacific Island peoples.

In what ES regarded as a great "enigma", two eminent scholars, W. H. Prescott and E. B. Tylor, had been inclined strongly towards diffusionism before backtracking. Prescott in his classic work *The Conquest of Mexico* 1843) had been much struck by analogies between New and Old World civilizations. He listed an impressive array of shared characteristics that seemed to show the Asian origins of Mesoamerican cultures: similar calendars, religious usages, pyramidal architecture, hieroglyphics, etc. Prescott noted that "the Aztec system of four great cycles, at the end of which the world was destroyed, to be again regenerated" was akin to belief in periodical convulsions of nature in "many countries in the eastern hemisphere". Mexican stories of the Flood were obviously similar to those of Hebrew and Chaldean narratives. The Catholic missionaries who first landed "in this world of wonder . . . were astonished by occasional glimpses of rites and ceremonies which reminded them of a purer faith . . . They could not suppress their wonder, as they beheld the Cross, the sacred emblem of their own faith, raised as an object of worship in the temples of Anahuac". The Aztecs had a form of communion similar to the eucharist ("an image of the tutelary deity of the Aztecs was made of the flour of the maize, mixed with blood, and, after consecration by the priests, was distributed among the people"). The probability of a cultural link with east Asia "is much strengthened by the resemblance of sacerdotal institutions, and of some religious rites, as those of marriage, and the burial of the dead; by the practice of human sacrifices, and even of cannibalism, traces of which are discernable in the Mongol races; and, lastly, by a conformity of social usages and manners, so striking, that the description of Montezuma's court may well pass for that of the Grand Khan's, as depicted by Maundeville and Marco Polo" (quotes from Prescott, pp. 89, 90, 91).

Prescott was cautious about inferring correspondence between nations from "a partial resemblance" of habits and institutions and he did not underplay the differences between Old and New World cultures. However he suggested that the accumulation of general coincidences "greatly strengthens the probability of a communication with the East" (quote, p. 92). He finished on a quite equivocal note:

> Whichever way we turn, the subject is full of embarrassment. It is easy, indeed, by fastening the attention on one portion of it, to come to a conclusion. In this way, while some feel little hesitation in pronouncing the American civilization original, others, no less

> certainly discern in it a Hebrew, or an Egyptian, or a Chinese, or a Tartar origin, as their eyes are attracted by the light of analogy too exclusively to this or that quarter. The number of contradictory lights, of itself, perplexes the judgement, and prevents us from arriving at a precise and positive inference. (quote, p. 95)

This was closer to the caution of an historian than to the desire for certainty of a scientist. ES, the scientist, felt that Prescott "evades the logical conclusion that seems so clearly to emerge from the evidence he cites and the interpretations of it he so emphatically expresses" (p. 96). He blamed Robertson's influence and Cartesian scholasticism for biasing Prescott's judgment.

Edward Burnett Tylor's case was even more puzzling, more of an enigma, than was that of Prescott, or so ES believed. Whereas Prescott had merely backtracked from an eminently rational position, Tylor managed most of his life to adhere to the confusingly different doctrines of diffusion and independent invention of culture. Criticizing Tylor was a hazardous business, as ES had already discovered. It brought wrath upon one's head. Tylor (1832–1917) was the acknowledged "father of anthropology", known for his definition of culture, his theories on the evolution of society and religion, his pioneering of statistical methods, his theories of animism and survivals. He had written famous books such as his *Early History of Mankind* (1865) and *Primitive Culture* (1871); he became the first professor of anthropology at Oxford (1896) and president of the Anthropological Society (1896), and was knighted in 1917. ES had enormous respect for Tylor's scientific work and lucid style, but he believed that Tylor had been led astray by an irrational adherence to his belief in animism, maintained long after it had been discredited by later ethnologists.

Tylor had carefully assembled an impressive array of facts, and used formidable arguments, to suggest that culture had been diffused between peoples. In his *Researches into the Early History of Mankind and the Development of Civilization* (1865) he presented evidence about practices such as ear-piercing, bodily deformations, folk-tales and myths, "games, popular sayings, superstitions, and the like" – all of which suggesting common origins in older cultures. He said of folk-tales and myths, for example, that "they show coincidences so quaint, so minute, and so complex, that they could hardly have arisen independently in two places" (quoted, p. 135). The story of the Flood and the Ark was one such instance. Tylor had a life-long interest in the history of games and wrote many scholarly papers on them, their geographical distribution, possible origins, etc. He pointed to the resemblances between dice-playing games such as backgammon, played by the ancient Romans, to games played in

Greece, Persia, ancient India (*pachisi*), Old Mexico (*patolitzli*), concluding that the Mexican game must have come from Asia (and ultimately the Middle East). He made studies also of games in the chess-group and the polo-hockey group, as well as kite-flying. He also found that the "Asiatic conception of the earth being arched over by a number of concentric heavens is found in Polynesian mythology . . . It seems quite unlikely that such notions of successive vaults or storeys above and below the earth should have sprung up as spontaneous fancies among the Polynesians, whereas they are quite explicable as borrowed from Asia, where ignorant priests had degraded them from astronomy to introduce them into religion" (quoted pp. 159, 161). Tylor spoke of the "drifting" of Asiatic culture, or peoples, into Polynesia.

Even closer to ES's heart, Tylor noted that the North American legend of a "Great Elk" "was a real reminiscence of a living proboscidean", a notion supported "by a remarkable drawing from one of the Mexican picture-writings". Both Tylor and ES agreed with the great naturalist (and early diffusionist) Alexander von Humboldt, who had first published this drawing, that it presented "some remarkable and apparently not accidental resemblances with the Hindoo Ganesa (the elephant-headed God of Wisdom)" (quoted, p. 138). On a triumphant note ES concluded:

> The evidence collected by Humboldt, d'Eichthal, Tylor, and many others, establishes the fact that the early civilization of America, its pyramids, its turbanned figures, its ear-plugs, its architectural ornaments, its folk-lore and mythology, are thoroughly Indian, or rather Indo-Chinese, in motive and feeling. Hence it is not surprising to find representations of the Indian elephant (with characteristically Indian conventionalization and additions to the designs) on the monuments and in the manuscripts. Obviously the admission of so conclusive a demonstration of Asiatic influence in America would be fatal to the pretence that the Maya civilization of Central America grew up without any outside help. (p. 141)

However, even in his early works, which were suffused with diffusionism and insisted on the importance of the historical method, Tylor had virtually excluded from this approach the prevalence of "a belief in the continuance of the soul's existence after death" (quoted, p. 170). Tylor argued that in this case independent invention seemed to prevail. In *Primitive Culture* and later works he developed this into a theory about animism. Objects such as stones and trees were widely believed by different peoples to be animated by spirits or souls. This seemed to be an instinctive belief. Here, he argued, were to be found the origins of religion. Tylor was also guilty, in ES's view, of the fallacies of "psychic unity" in humankind and belief in a general uniformity in evolutionary stages (the "stage theory" ES had always attacked as basically un-Darwinian).

ES commented: "Hence we have the strange phenomenon of a persuasive writer of transparent sincerity and exceptional authority maintaining throughout thirty years of his life the case for diffusion of culture at the same time that he was the leading exponent of the diametrically opposite interpretation of human action which is commonly called the independent development of culture" (p. 174). Even stranger, the scientific methodology that Tylor espoused resulted in later scholars doing intensive studies of animism, revealing that animism was not universal, but limited "both in character, in geographical distribution, and in time . . . The demolition of this doctrine, however, did not seem to affect the theories that were founded upon it. Those who drew their inspiration from Tylor still retained their childlike faith in the reality of the independent development of culture" (p. 175).

Whether this was good history of science is problematic, to say the least. ES in fact showed an excellent sense of historical context when he explained the genesis of Tylor's scientific paradigm in anthropology in the context of the exciting intellectual ferment of the 1860s – the world of Darwin's *Origin of Species*, Lyell's revolutionary geology and principle of continuity, the work of Dalton and Faraday in physics, of *Essays and Reviews* in theology, of George Meredith and George Eliot in literature. He speculated that Tylor's early Quakerism and fear of secularism may have conditioned his theory of animism. But, in his own time, ES was inclined to see polarities when there were complexities. He stubbornly ignored the fact that, as Malinowski observed, the early evolutionists commonly combined diffusionist ideas with those of independent invention. He tended to identify independent invention with the psychological theory of the "psychic unity" of humankind, but failed to see that the latter theory – however contentious – was not necessarily inconsistent with diffusionist ideas.[17]

ES's frame of mind is exemplified by his summing up of the significance of the work of Robertson, Prescott and Tylor. Their writings illustrated "the growth of the strange superstition of independent development of culture". The curious fact emerged "that when the possibility of the diffusion of culture was being seriously considered, as it was by Prescott and Tylor, precise and detailed evidence was cited to establish its reality: yet when the alternative hypothesis of independent development was being advocated, the actual evidence was ignored and resort was made to unsupported speculation of the idlest and most irrelevant kind". In his view, the doctrine of independent invention, freed from "the hampering influence of reliance upon evidence, took on a new lease of life, and developed into the luxuriant growth of wild speculations that for more than forty years has flourished in the whole field of ethnology. The Newtonian discipline was abandoned and Cartesian methods of a priory deduction became dominant" (pp. 178–179).

Needless to say, ES reiterated his theory of Egypt and cultural diffusion, repeating much of what he had said in earlier works, most recently *Human History*. The dense detail of works such as *The Evolution of the Dragon* was now replaced by more flowing prose that could be understood more easily by a general audience (although he still could not resist chasing hares up long alleys on topics that fascinated him – this gave a rather unbalanced feel to the book). He felt vindicated by ongoing research, especially by scholars such as Thomas Cherry and C. Daryll Forde. Cherry had put new evidence that Egypt had first invented agriculture and irrigation. The inundation of the Nile stimulated the first measurement of the year and was responsible (Cherry argued) for the creation of the calendar. Daryll Forde's book *Ancient Mariners* (1927) concluded that Egypt invented sea-going ships, and that vessels of the Egyptian type had encircled the world.

One striking feature of ES's last book was an emerging enthusiasm for Islam. He had long cited the great religions of the world – Christianity, Islam, Buddhism – as examples of cultural diffusion. Each of them had spread from a single location. Each had been created by a single individual:

> Each was adopted by the most diverse peoples, irrespective of race, circumstances, and degrees of culture. These religions in different degrees assimilated the most varied accretions in the course of their diffusion, and became so profoundly modified that in some cases little resemblance to their original character remained. At the same time, they influenced every aspect of the customs and ideas of the peoples who adopted them, their arts and crafts, in particular their architecture and symbolism, their manner of life, their social customs of marriage, eating and drinking, and their moral principles – in short, their whole civilization. They display not only the reality of diffusion, but also the factors that determine its practice. (p. 12)

In his quest to explore further such dynamics, ES had begun reading widely on Islam, including books such as Ameer Ali's *The Spirit of Islam* (1902) and De Lacy O'Leary's *Arabic Thought and its Place in History* (1922). These so captured his imagination that he rather unbalanced the introduction to his book by devoting 22 out of 37 pages to a detailed analysis of Islam. (This was something that he had often done in his writings, becoming so obsessed with a particular topic that he could not resist writing about it at great length, often straying wildly from his main thesis.) His main point was that Islam had spread so widely and successfully[18] not just because it was a profound religion, not just because of its passion for knowledge, its science and scholarship, but because of its enlightened and essentially tolerant spirit. This was not necessarily the prevailing image of Islam in the west in 1933, just as it is not so now. He pictured Islam being diffused (like Christianity) within a few centuries from western Asia to the Atlantic and the Pacific littorals. However, unlike some forms of puritanical Christianity, Islam was adaptable, elastic and liberal: "The radical changes introduced into the body spiritual and politic of Islam through full contact with the heterogeneous cultures of the civilized world not only acquit it of inelasticity, but also throw fresh light on the way diffusion works (p. 21) . . . As marked and universal a cause of the spread of Islam was its genius for toleration, the primal impulse of which may have come from the Persia of the Achaeminids (p. 24)".

His focus was upon mainstream Islam, not belligerent minor sects or ideologies. He blamed biased Christian historians for focusing on the latter. One such historian, for example, had declared: "Every Musulman who fell in the warfare to found a universal monarchy was promised a residence in the society of the black-eyed houris". This, said ES, was a crude judgment that took no account of the obvious truth that coercion enforced hypocrisy not conviction, and without glad conviction the broad diffusion of Islam would have been impossible:

> As for the black-eyed houris, their delectable society represents as faithful a reflection of the meaning of Islam as the Court of Charles II does that of Christianity. Islam was quite definitely not propagated by the sword, neither were its wars invariably barbarous or internecine. (pp. 25–26)

Even (or especially) in the post 9/11 world of jihadists and suicide bombers, ES's reminder about mainline Islam is timely.

ES could not resist a final swipe at his Malinowskian critics, who persisted in stereotyping his theory as "mechanistic":

> The diffusion of culture is not a mere mechanical process such as the simple exchange of material objects. It is a vital process involving the unpredictable behaviour of the human beings who are the transmitters and those who are the receivers of the borrowed and inevitably modified elements of culture. Of the ideas and information submitted to any individual only parts are adopted: the choice is determined by the personal feelings and circumstances of the receiver. Moreover, the borrowed ideas become integrated into

the receiver's personality and more or less modified in the process of adaptation to his knowledge and interests. Such selection and transformation occur in all diffusion of culture not only from one individual to another, but even more profoundly in the passage from one community to another. The vehicles of transmission are affective human beings, and subtle changes are introduced into every cultural exchange in accordance with their personal likes and dislikes, no less than in their ability and understanding and the circumstances at particular moments. (p. 10)

CHAPTER

7

Last Days (1933–1937)

Life began to go awry for ES from about the time of his wife's illness in October 1931. She had a serious bout of broncho-pneumonia that hospitalized her and lasted for months. He declared himself "desperately worried". His workload had been, as usual, onerous. There had been an arduous journey to China in 1930, trips to Copenhagen and Madrid in 1931, a congress of Prehistoric and Protohistoric Sciences in London in 1932, endless public addresses and conferences, even radio broadcasts. He was contributing a chapter to a book on the neural basis of thought, another to a textbook of anatomy, and was completing his *Diffusion of Culture*. His health, usually robust, broke down in the winter of 1932. According to Warren Dawson: "In December, whilst returning from Cambridge, where he had been examining, he had a seizure. After some months in hospital and convalescence at Broadstairs, he made a marvelous recovery, and although suffering from physical disability, he was mentally alert as ever, and soon immersed himself in his old interests".[1]

ES's young friend Solly Zuckerman, later an eminent zoologist and influential adviser to British governments, remembered the events a little differently. ES had helped Zuckerman to launch a career in the Zoological Society of London. Late in 1932, ES asked Zuckerman to stand in for him and give a paper on Peking Man to the Royal Anthropological Institute. Zuckerman dissented from ES's classification of hominids and proposed a complete revision of the classification. He showed his paper to ES, expecting opposition. Instead ES actively encouraged Zuckerman to publish his paper. Zuckerman duly gave his talk:

> When we left the Anthropological Institute, just off Gower Street, Elliot Smith asked whether I would accompany him home. This was in the days before many people had cars, and we had to change buses at Camden Town in order to get to his house in Albert Street. While we were waiting for our second bus, he took out his hand-

> kerchief, and as he wiped the side of his nose, he remarked strangely, 'I am wondering which of my lenticulo-striate vessels are leaking. All day I have felt as though I've a cold on the right side of my nose, and there's been a slight tingling in my fingers. I wonder how serious a stroke it's going to be'. I was more than a little shaken, but he seemed as calm as could be. I took him home, told his wife that he wasn't well, and called next morning – I did not live far away – to see how he was. Two or three days later, he was in University College Hospital following a stroke, from the effects of which he never fully recovered.[2]

ES's stroke may have been mild, but it was enough to debar him from full activities at University College during 1933. He was temporarily replaced by H. A. Harris who would later become permanent head of anatomy. Harris commented: "Many were able to watch the struggle which Elliot Smith made during his illness from 1932 to 1934, courageously attempting to perform his professorial duties. A few were able to see his repeated efforts at staging a comeback during his retirement from 1934 to 1936".[3] In February 1934 he was diagnosed with diabetes and was hospitalized at University College for a month.

Much of ES's waning energy was devoted to a campaign against what he believed was the odious Aryan race theory being pedaled by the new Nazi regime in Germany. Liberal academics in Britain, including anthropologists, were alarmed and activated by the Third Reich's agenda of anti-Semitism, eugenics ("racial hygiene"), totalitarian control of education and suppression of academic freedoms. Moves were made for the anthropological profession to make a strong statement against the "pseudo-science" of Nazi race theory. However no oppositional consensus to the Nazi agenda was reached by anthropology, or academe generally. The Royal Anthropological Institute set up a Race and Culture committee in 1934 to offer a rigorously scientific definition of race. Elliot Smith and his liberal colleagues hoped for a striking result. But the experts failed to agree and the report was a damp squib, later criticized as a "failure of nerve". This was deeply disappointing to ES, but it was explicable as the result of personal and professional rivalries, fears about the "politicization" of the discipline, differing readings of events and strategies, and epistemic differences over the admittedly problematic area of "race". I have examined the history of the Race and Culture Committee in detail elsewhere, and the interested reader can find full documentation there.[4] However I will briefly recount here ES's role in the events of 1934–1936.

In 1932 ES had been appointed founding chairman of the Human Biology group within the Royal Anthropological Institute (RAI).[5] He had kept at arms length from the RAI (Royal Anthropological Institute) for some years, partly because it refused to modernize its scientific methodology, but also, it seems, because his old foe J. L. Myres ran the show (Myres was president of the RAI from 1928 to 1931 and was editor of the leading anthropological journal *Man* from 1931 to 1946). However ES agreed to chair the new committee because he championed a biologically-based rejuvenation for anthropology. He resented Myres's lack of commitment to this goal, accusing him (in a letter to C. G. Seligman) of "chronic cussedness".[6] In his founding discussion paper on the purpose of the Human Biology committee, ES complained that British anthropology was out of touch with biology. As a result "there is no ignoring the fact that within recent years biologists have neglected the Institute and left the way open for more or less sensational popular expositions of the more hackneyed aspects of physical anthropology".[7]

ES organized a joint Human Biology and Sociology meeting at the RAI on 24 April 1934. It recommended to the RAI council the setting up of a "Committee for Race and Culture" to discuss "The Significance of the Racial Factor as a Basis in Cultural Development". This was a response to a complex of factors, including alarm from European scholars about rising anti-Semitism and sympathy for Jewish intellectuals fleeing the Nazis. The Czech physician Ignaz Zollschan, a Jew, came to Britain at this time and played a key role in orchestrating an anti-racist campaign. His contacts included the major anthropologists in England. Perhaps most important was Charles Seligman, the Jewish ethnologist, head of anthropology at LSE until 1930, who played a key role in having the committee set up. (However he and other Jewish scholars were formally excluded from the committee on the grounds that this would look more "objective" – in fact Seligman still played an unofficial role as an adviser.) The leading American anthropologist Franz Boas, a friend of ES, was also very active behind the scenes. ES kept in close contact with Boas (as is shown in their correspondence). Malinowski, who was anti-Nazi, gave general encouragement to the project, but did not participate. Myres also facilitated Zollschan's work, although he found him infuriatingly unpredictable. A key aim of the activists was to get a strong anti-Nazi motion passed at the International Congress of Anthropological and Ethnological Science, to be held in London in July 1934.

ES agreed to chair the Race and Culture committee, and Raymond Firth was appointed honorary secretary. When ES's illness worsened, the geographer H. J. Fleure, a friend of Myres, became acting chairman. This did not augur well, as Myres was very ambivalent about the project, worrying about causing divisions within the profession. The committee was diverse, ranging from anthropology and geography to biometrics and biology, from "liberals" like ES and Le Gros Clark to arch-conservatives like Reginald Ruggles Gates and G. H. L. F. Pitt-Rivers (a late appointee). It took two years to complete a report.

The RAI responded to concerns from Myres and others by adopting a policy of "no publicity" about the committee. It stubbornly stuck to this policy. ES however spilt the beans about the committee in a letter to the *Times* which protested angrily about German prohibitions upon Jewish-German marriages.[8] Myres was furious. He marginalized against his clipping from the *Times*: "I warned the *Times* that the R. A. I. enquiry was confidential, and expressed the hope that there would be no further comment".[9]

The July congress proved something of a fiasco for ES and his friends. As chairman of section A of the congress (Physical Anthropology), ES gave a provocative opening address. As recalled by the anatomist A. A. Abbie: "There was a strong delegation of German anthropologists at the meeting, and, in his opening address, Elliot Smith pointedly demolished any claims to a 'Nordic Race' or an 'Aryan People' . . . The German delegates looked extremely uncomfortable but made no protest – they were in a very delicate position! At that Congress, too, it was evident that he [ES] had lost none of his adroitness in handling either difficult questions or troublesome people".[10] Cheers greeted his quotation from Max Müller that "an ethnologist who speaks of Aryan race, Aryan blood, Aryan eyes and hair is as great a sinner as the linguist who speaks of a dolichocephalic dictionary or a brachhycephalic grammar". Aryan and Semitic were linguistic terms, and did not imply separate races. ES declared: "we in this section are surely within our rights in criticizing fallacies that come into flagrant conflict with the generally recognized teaching of anthropological science".[11] The *New York Times* headlined its feature "Scientist Assails 'Aryan Fallacy'" and heralded his speech as "the most outspoken that has been heard at a British scientific meeting in a long time".[12] The British press reacted with characteristic restraint. Myres, astonishingly, made no mention at all of ES's address in his otherwise detailed account of the congress published in *Man*.

It was all too "political". Virtually nothing of a controversial nature in fact found its way into print in *Man* under Myres's editorship in the 1930s, and the same was true of publications emanating from the RAI.

Behind the scenes at the congress, moves were successfully made to head off any embarrassing "political" statement on race. ES's correspondence with Seligman reveals the intrigues that were going on. It appeared that the executive bureau of the congress, of which Myres was secretary, had blocked a draft memo "Remarks on Race", written by ES. He retaliated by going public. He arranged an interview with the *Observer*, writing to Seligman that "the Bureau must be prevented from such Hitlerite abuse of power".[13] The interview caused a stir, and evoked much public support. Many anthropologists wrote to him expressing their agreement with his action. However another draft resolution on race was again blocked by those in charge. According to ES the powers-that-be considered it too "dangerous". Although the *New York Times* again splashed what had happened, the whole affair has been almost completely covered up in the historical record. Practically nothing was reported in the scientific journals about the suppression of what was by no means a radical resolution on a major contemporary issue.

As 1934 wore on ES, Zollschan and others continued to press for a statement on race from British anthropologists, but to no avail. ES asked Boas to send Raymond Firth thirty copies of his pamphlet *Aryans and Non-Aryans*, a direct attack upon Nazi race theory: "I want all members [of the Race and Culture committee] to read your pamphlet before we draft our final report".[14] On 11 December 1934 the committee gathered to discuss eight draft memos on race, including one by ES. As meetings and further draft circulations on race followed each other during 1935, Myres opposed any leaks to the press.

ES gave a strong indication of his views in an address to the Rationalist Society entitled "The Aryan Question", in which he reviewed the history of "this monstrous fallacy". He concluded:

> Bearing in mind the considerations discussed in my book *The Diffusion of Culture*, we can appreciate something of the fallacy involved in the claim that the Aryans alone among the peoples of the world have been founders of culture. This is merely a revival of the claims made in the middle of last century for the Nordic race by the Comte de Gobineau . . . We have a sufficient mass of exact evidence today to be sure that the claims made for the Nordic strain are devoid of any adequate justification. The particular culture a people possesses depends mainly upon historical circumstances, and not upon any hypothetical innate qualities that may be attributed to them. While frankly admitting that the Indo-European language must originally have been developed among one particular people which might be called the primitive Aryan race, it is equally certain that there is no race living at the present time which can truly be called 'Aryan'.[15]

ES was by now desperately ill and in no position to dragoon the committee into a robust protest against the Nazis (even if that had been possible). Myres, afraid that "the stunt-factories" would have it all their own way over the committee's activities, kept the lid on proceedings.[16] Meantime the liberal-minded biologist Julian Huxley pre-empted matters with a vigorous attack upon totalitarian race theory in his popular book *We Europeans* (1935).[17] Huxley's book was freely available to the committee. Remarkably, it was completely ignored in the report.

That report, entitled "Race and Culture", finally appeared in April 1936 as a twenty-four page pamphlet published by Le Play House press. It failed to find a scientific consensus, was full of contradictions, and satisfied virtually nobody. It received very little press coverage, and was quickly forgotten. Unable to reach agreement on central issues, the committee offered alternative definitions of race, followed by separate comments by seven of the panel of twelve (Myres and four others offered no opinions). Ruggles Gates continued to defend his outdated theory that major races, such as white, black and yellow, were separate species. Fleure and ES disputed this, the others pushed their own specialist interests. Pitt-Rivers warned about the perils of politicizing definitions that should be strictly scientific. ES repeated the views he had expressed at the congress and in his various papers. He ridiculed genetically determinist theories about culture and race: "What little evidence we have of the history of culture . . . seems definitely to establish the fact that the acquisition of culture is not due so much to innate qualities as to historical circumstances and quite arbitrary factors".[18] This was similar to Gordon Childe's stance that culture was independent of physical race, and was largely a matter of social tradition.

The RAI rested on its oars. Something at least could be said to have been done. When the Second World War broke out, the physical anthropologist Geoffrey Morant, a member of the committee, reflected that the dismal outcome of the Race and Culture inquiry was just one symptom of the glaring failure of European anthropologists to counter the racial fallacies propagated by the Nazis. In an address to the RAI, he noted sarcastically that the committee's pamphlet would probably not have come to the notice of Dr Goebbels, "and we must admit that it would not have caused him much anxiety if he had seen it". He observed that ideas about race had become conditioned by nationalist cults but "anthropologists made no effective protest when these matters became of prime importance to the peace of the world".[19]

ES retired from University College at the end of the 1935–36 academic session (June 1936). As news of his illness spread, and he was obliged to withdraw from his duties, he was showered with honours. He had won many distinctions over his life: elected Fellow of the Royal Society in 1907, the prized Prix Fauvelle from the Anthropological Society of Paris in 1911, a Royal Medal in 1912, a Gold Medal from the Royal College of surgeons in 1930, and various honorary doctorates and memberships of scholarly societies across the world. He was gratified to be elected an Honorary Fellow of St John's College, Cambridge, in 1931. Then in 1934 he was granted a knighthood, while in 1936 he was awarded the Decoration of Chevalier de l'Ordre National de la Légion d'Honneur. The RAI belatedly recognized him by awarding him the Huxley Medal in 1935 and electing him Huxley Memorial Lecturer. In that lecture – read on his behalf by a colleague in November – he heaped praise on his hero T. H. Huxley. Huxley, he said, had directly inspired his interest in analyzing the brain structure of placental mammals, which led ultimately to his involvement in anthropology. Huxley, a key founder of the RAI, had been one of the first to insist upon the biological foundations of the study of humankind.[20]

As Warren Dawson remembered:

> Early in 1936, Elliot Smith's health suffered a severe setback in consequence of two domestic tragedies: Soon after Christmas [1935], his youngest son [Stephen] lost his life as a result of an accident, and a short time afterwards Lady Elliot Smith had a serious accident which confined her for many months to hospital. Whilst deeply moved, Elliot Smith submitted stoically and uncomplainingly to his fate, and his peace of mind was further disturbed by the fact that the lease of his house had expired, and he was required to vacate it immediately as it was about to be demolished. As Lady Elliot Smith was still in hospital, he entered a nursing-home at Sidcup, where he remained until a few days before his death.[21]

ES tended to lease rather than own houses. He had not in fact earned much wealth from his job and his publications (his estate on his death was valued at just over £4600). He had leased his house in Albert Street before his stroke in 1932. As he wrote to the secretary of the Zoological Club, a venerable dining club belonging to the Zoological Society, a favourite haunt of his because the cream of the scientific world gathered there: "Two years ago my chief reason for choosing this house to live in was its proximity to the Zoo and the Zoological Club, but immediately after moving in I was taken ill and that defeated the very purpose of the move".[22] Late in October his wife went to St Michael's nursing home at Broadstairs, Kent. At Sidcup ES kept in touch with his friends and affairs, and began an autobiography. On Christmas Eve he joined Kate at Broadstairs. But as Dawson recalled: "the journey and the excitement were too much for his weakened health" (p. 110). Grafton Elliot Smith died on New Year's Day, 1937, aged 65.

Afterword

The diffusion controversy has never really ceased, but it has continued to be played in a minor key whereas in Elliot Smith's time it was in major key. To change metaphors, since the thirties it has lived a subterranean existence, breaking to the surface now and then as part of larger theoretical debates. Diffusionism was once part of a ruling paradigm in anthropology and archaeology, but, as we have seen, it was supplanted by new ruling paradigms. As happens in science, the practitioners of new paradigms take over from the old school, gradually (sometimes rapidly) marginalizing them, even demonizing and excluding them. This was happening during the latter years of Elliot Smith's lifetime, but accelerated after his dominating presence had gone.

The generation of students after the Second World War was never exposed to the writings and teachings of the Elliot Smith–Perry school. Their books disappeared from the curricula, even from academic libraries. In an illuminating comment, the Texan geographer G. F. Carter wrote to a colleague in the 1970s: "I never read Grafton Elliot Smith, for I was properly raised in anthropology (A. B., U.C., Berkeley) and knew that he was crazy. One doesn't waste time reading mad men!" But, as he had gotten an interest ethnologically in the symbolism of elephants and eventually read ES's *Elephants and Ethnologists*, "I met a witty, urbane, informed, delightful person". Carter then added his opinion that "GES's treatment is a scientific and scholarly disgrace".[1]

Leading the charge against diffusionism in the 1960s was the eminent prehistorian Glyn Daniel. His book *The Idea of Prehistory* (1962) contained an influential critique of Smith–Perry "hyperdiffusionist doctrines". Its language was inflammatory – Daniel freely used terms such as "academic rubbish", declared that ES and Perry had "really abandoned any pretence at scientific method" and consigned them to the "lunatic fringe" of prehistory, along with believers in the Lost Tribes of Israel and Atlantis.[2] Daniel (1914–1986) came from relatively humble origins in Wales, the son of a local schoolmaster. By sheer talent he made a brilliant career for himself at St John's College, Cambridge (his ashes are scattered in the college gardens), becoming an expert on megalithic monuments of western Europe. He wrote popular books on the themes and history of archaeology, became editor of the key journal *Antiquity* (in 1958) and a leading television personality (famous as chairman of the panel game "Animal, Vegetable, Mineral?"). Not unlike ES's other foe, John Linton Myres, Daniel became a mandarin in the archaeology/anthropology profession.[3]

According to Daniel, the evidence supported his strong belief that the first civilization began, not in Egypt, but in Mesopotamia, with the Sumerians. ES had thought that this was a possibility, but, in the years before carbon-dating, he lamented that exact chronologies were not available (and he insisted that they were desperately needed). Daniel argued that older theories such as unilinear evolution, or diffusionism (certainly the extreme varieties), were now outmoded. In books such as *The First Civilizations* (1968) he championed multilinear evolution, leading inevitably "for some societies with geographical and ecological and cultural possibilities" to lay down civilizations featuring early urbanization and so on. Following in the steps of scholars such as Julian Stewart, A. L. Kroeber and Joseph R. Caldwell, Daniel postulated that civilization was an inevitable response to laws governing the growth of culture, the result of a finite number of social and historical processes: "the lesson of archaeology at the present day", he wrote in 1968, "is, I suggest, that seven societies in seven different ways trod one of these paths, the path that led to civilization".[4]

ES would have gone along with much of this. He also believed that civilization arose out of a limited number of cultural, geographical and economic factors. While sceptical about Mesoamerica developing independently, he agreed that Indian and Chinese cultures had evolved largely indigenously, although external influences had also been important, providing key stimuli at key historical moments. Daniel himself admitted that diffusion occurred, although he circumscribed it much more severely than did ES. It was one thing, he said, "to reject as the archeological evidence tells us to reject, the idea of civilization coming to America from outside; it is quite another thing to deny the possibility of influences coming to America from outside". He conceded that sea travel was possible across the Pacific to America. And there were some interesting parallels between American and Asiatic artifacts. Gordon Ekholm had documented close resemblances between the wheeled toys of Central America and some Asian toys. Meggers, Evans and Estrada had found exact parallels between pottery unearthed on the coast of Equador and some Neolithic pottery in Japan. Granted this, "There is nowadays no suggestion of any direct transplantation of an Old World civilization to the New, and indeed no suggestion of Old World–New World contacts comparable to the Mesopotamian–Egyptian catalysis . . . Any contacts trans-Atlantic or trans-Pacific that may have occurred were slight and very infrequent and had little effect on the native development of pre-Columbian American culture".[5]

Daniel did not share the pessimism about human creativity that ES certainly displayed most of the time (and also Gordon Childe, a founder of the new "cultural-historical" school of archaelology. Childe and ES had much in common, apart from the fact that both were Australian. Not least was their emphasis on diffusion and migration as agents of cultural change. Daniel felt that a modified diffusionist paradigm, associated with Gordon Childe and Daryll Forde, had dominated European prehistory from the 1930s to the 1950s, and that it had sprung up in reaction to ES's "hyperdiffusionism"). Daniel insisted that "savages" did invent many things (for example, agriculture, the domestication of animals), and he believed that independent invention was common. The barbarian agriculturist invented civilization in southern Mesopotamia: "So it is no longer any use saying that the savage invented nothing and the barbarian invented nothing . . . And it is just not true to say that things cannot be discovered or invented more than once; it is nonsense and results from a refusal to study the history of invention and discovery in all its realms – prehistory, ethnology, and history itself . . . It is equally at variance with the facts to say that stray cultural contacts cannot occur, and that ideas and techniques cannot be diffused in a simple and individual way". Joseph Needham's research had demonstrated the diffusion of mechanical and other techniques from China to the west. Therefore "we should not boggle at the possibility in earlier times of wheat-growing, the alloying of copper and tin to make bronze, and the *cire-perdue* process of bronze-casting coming from west to east, or rather from the most ancient Near East to China". However the single-origin explanation of civilizations was too simple – such processes were invariably complex.[6]

ES's ideas had certainly lost credence in the professional establishment by the 1970s. This was demonstrated, ironically, in a symposium organized in his honour in November 1972, roughly the centenary of his birth. The symposium was arranged jointly by the British Anatomical and Zoological Societies, and was held at one of ES's favourite haunts, The Zoological Society headquarters in London. His old friend Solly Zuckerman (now Lord Zuckerman) was organizer and edited the proceedings for publication.[7] He was generous in his praise of ES and so were most of those who spoke about his medical and scientific achievements. But the tenor was notably different in the session on his prehistoric archaeology and cultural anthropology, a session that was chaired by Glyn Daniel. This dismayed latter-day diffusionists: "It was like having asked Russia's Communist Party chief Brezhnev to preside over a meeting in honour of the dissident author Solzhenitzyn". The "new diffusionists" were allowed only a very restricted role during the session.[8] Unflattering adjectives such as "extreme", "delusional", "dogmatic" and "fallacious" were to be heard, and there was a whiff of heresy-hunting in the air. Be warned, it was said, there are still hyper-diffusionists among us.

In the event Glyn Daniel's long address was a serious critique of ES's ideas. Occasionally harsh, it was less polemical than diffusionists may have expected. Daniel praised ES for being the first to pursue the Egyptian hypothesis – one that had been around from the time of Herodotus – with exact archaeological parallels: "His was the first clear statement of diffusion from one area in archaeological and anthropological terms". However, archaeological research, including exact dating of artifacts, had moved the whole debate into new realms:

> Only the most heavily blinkered diffusionists – and they still exist – would now deny that what we conceptualize as agriculture was independently developed in many parts of the world, such as the Near East, China, and Central America, and that what we group together mistakenly as megalithic monuments represent a wide variety of different structures separated in time, space and origins. But we should remember here in fairness that Elliot Smith was writing years before Libby discovered Carbon 14 dating . . . [1963], and although he based his case extensively on the origins of agriculture and megaliths, he did not have at his disposal the detailed and exact knowledge that we now have.[9]

According to Daniel, ES had never effectively documented a crucial connection between Egyptian mastabas and megalithic tombs in Europe. When he failed to find transitional types, he quite unacceptably hypothesized them.[10] More generally Daniel felt that both the hyperdiffusionism of ES (indeed all monocentric systems) and the modified diffusionism of Childe were no longer suitable paradigms: "Both were useful in their time, Elliot Smith's Egyptian paradigm was a useful one, and its extremes and extravagancies played a useful and important part in the development of theories of cultural change" (p. 419). However, obstinately heeding no criticism, ES had "acquired a set attitude of mind with regard to the Egyptian origin of all culture and civilization, and this *idée fixe* had become, sadly, a delusion" (p. 419). As we have seen in an earlier chapter, Solly Zuckermann strongly defended ES against any charge of dogmatism and praised his essentially scientific spirit of enquiry. Zuckerman finished own his peroration with the comment:

> His own opinions, right or wrong, were founded on facts which were open to all. Equally, and to the best of my judgement, he

would always have abandoned views which he may have held when persuaded by evidence that they were wrong. The more he fought for his views, and however heated the argument, the brighter the light that was shed, and the wider the horizon became. What more can be asked of any scientist? (p. 20)

However Daniel's criticism was echoed by others, including Darryl Forde (who died six months later) and the Cambridge social anthropologist Meyer Fortes. Forde made the point that there was little possibility of discussion in anthropology between people like Rivers and ES on one side, and Malinowski on the other, "not because either of them was necessarily attacking the other's views, but because they were working on entirely separate lines" (p. 426). Fortes added the insight that ES's diffusionist scheme "was not fitted to formulate questions or suggest ways of answering questions that were relevant for the synchronic functionalist enquiries". It seemed a very interesting point to him "that Elliot Smith and Rivers were medically trained biologists by profession originally, whereas Malinowski was a physicist and therefore trained, I suppose, to think in terms of the constancy of natural systems and not in terms of evolutionary chronology" (pp. 431–432). Why was ES so single-minded in his espousal of unicentric diffusionism? Fortes pinpointed his "all-embracing commitment to 19th century, Darwinian evolutionary biology . . . This Egyptian-centred theory of diffusion, I would suggest, fits very well into a monogenetic Darwinian type of biological evolutionism – a conception of an evolutionary tree – and a habit of mind which prompts one always to seek explanations for contemporary states of affairs in terms not merely of origins, but of remote and ultimate origins" (pp. 428–429).

Fortes accused ES of stripping away the facts of culture from primordial mankind: "It was constructing a picture of mankind before mankind had a history, which meant that aside from speech, fire and a few elementary tools, human nature was a complete cultural blank in the beginning . . . This reductionist procedure practically compelled him to resort to a picture of what amounted to a Golden Age of humanity supposed to be reflected in the origin myths and stories of the simplest living peoples" (p. 429). After two world wars and the Holocaust, Fortes was scathing about this Arcadian view of human nature. However he believed that diffusionism had continuing relevance (even if the Egyptian model had been discredited), since "it did assume the possibility of developing a unified view of the sciences of man, or, if not unified, a view of the sciences of man in their widest extent, which stresses their inter-connexion with one another and seeks to make use of their mutual complementarities, in a context combining historical perspective with synchronic analysis" (p. 432).

The diffusionists present tried to hit back in defence of ES. The Mesoamerican archaeologist R. A. Jairazbhoy claimed to have discovered a hundred or more evidences of Egyptian contact within the recently unearthed Olmec culture. This must, he felt, be a great blow "to the isolationists who have been advocating the independent origins of American civilization" (p. 442). Gerhard Kraus and C. E. Joel claimed that scholars such as C. L. Riley and others now accepted at least the possibility of Old World influence in any examination of the issue of American cultural origins.[11] Kraus and Joel admitted that archaeological evidence generally discovered since ES's death had "complicated the picture". Nevertheless, this did not mean that "Elliot Smith's great vision of the origin and dissemination of civilization" could be written off. For instance, Egypt may not have been the original starting point for agriculture: "Food production may have begun earlier independently of Egypt but it was there that cultivation and animal husbandry eventually flourished, fostered the growth of the ancillary features of early civilization and set the pattern for the rest of the world". They pointed out that ES's work on mummification still stood up, referring in particular to Graeme Pretty's re-evaluation in 1963 of the Macleay Museum mummy from the Torres Strait (which we have already encountered). Although ES's ideas were widely condemned as being out of touch, the reality was that many of them had still not been refuted, "and they have strong claims to be regarded as potentially valuable lines of approach to still unsolved problems" (pp. 442–446).[12]

The latter day diffusionists were resisting the central point that was made at the symposium, namely, that unicentric diffusionism was obsolete, that the whole debate had moved on. However there was a sub-text that emerged also. This was that the diffusionists and the schools that followed them, such as functionalism, occupied different worlds of discourse, speaking different languages. They did not in practice engage with each other. Much of the evidence and thinking produced by people like ES remained largely unchallenged. Different theoretical models, requiring different data, ruled the day, and preoccupied the attention of most scholars.

Alarmed by this situation, Kraus and Joel had set up their own journal in 1970. Based in England, it was called *The New Diffusionist*. It changed its name in 1979 to *Historical Diffusionism*, and was to last until 1981. It provided a forum for debate about diffusionism and over the years gathered together a mass of evidence and information about cross-cultural contacts, much of it about pre-Columbian contacts between the old and new world hemispheres. Important revisionist work was being done from the late 1960s onward in this respect, taking the simpler diffusionist models of the 1920 and 1930s into a more sophisticated theoretical context and employing new technologies in areas such as ceramic analysis, plant and animal distributions and micro-biology. Thor Heyerdahl's headline-grabbing voyage from Peru to Raroia in eastern Polynesia on the balsa-raft *Kon-Tiki* in 1947 had raised new interest in trans-Pacific voyaging, as did his following trip in the Egyptian-style reed-bundle craft *Ra II* in 1970. Trans-Atlantic crossings were shown to be possible when a scientific team led by Tim Severin crossed from Ireland to Newfoundland in a hide-covered wooden boat in 1977. (And, to add a very recent note, genetic researchers have just found that a woman from the Americas probably arrived in Iceland 1,000 years ago, leaving genes that have been detected in about eighty Icelanders today.[13]) Asiatic-American contacts were being increasingly studied from the early 1950s. Gordon Ekholm (an expert on wheeled toys in Mexico) suggested a possible Chinese origin of Teotihuacán cylindrical tripod pottery. The German scholar Robert Heine-Geldern did the same for the pottery of Mexico, Central America and Columbia. Working together the two scholars found significant parallels in the symbolic arts of southern Asia and Middle America.[14] ES of course had postulated such parallels.

American archaeologists and anthropologists generally remained stubbornly opposed to such ideas (although the resistance varied within sub-disciplines, it should be said). Why the intransigence? Various explanations have been advanced. They range from genuine epistemic concerns to alleged factors such as ingrained conservatism; territorial fears with a felt need to protect reputations and careers; the rule of an elder generation, with fixed mind-sets derived from youth and professional education, who controlled the destiny of younger scholars with more open minds and greater receptivity to new technology,[15] to factors such as parochialism and misguided nationalism (ES's old "Monroe Doctrine" mind-set). From an objective perspective, of course, it should have mattered not a scrap to the American sense of identity that the older Mesoamerican civilizations may have borrowed from other cultures. After all, the present American population is derived from a "melting pot" of peoples who "settled" or "invaded" the territories of indigenous groups, and brought with them the cultural baggage of their homelands.

However history of science studies shows us that scientific theory has always been conditioned by "non-epistemic" factors. They include nationalism, politics, racial and other societal values. If this applies even to supposedly "hard sciences" such as biology,[16] it applies even more forcefully to disciplines such as archaeology and anthropology. (Its practitioners, unfortunately, are given insufficient education on the history of their discipline, although postprocessualists rightly take account of subjective factors.) We need, of course, to analyse ES's ideas in this light.

But so too we need to scrutinize the often tunnel-visioned world-view of the staunch anti-diffusionists.

Although most of the participants in Solly Zuckerman's conference commemorating ES were blissfully unaware of it, a significant symposium on pre-Columbian contacts had taken place at Santa Fe in 1968. It was arranged by the Society for American Archaeology. Over thirty scholars discussed the state of the diffusion controversy, cultural contacts *within* the New World (even this was stlll being denied by many main stream academics), possible trans-Pacific and trans-Atlantic contacts, and more technical issues such as plant dispersal around the world. Although opinions were diverse, some extremely interesting evidence emerged. For instance, the anthropologist John L. Sorenson had for fifteen years been assiduously collecting a corpus of evidence from the literature that concerned connections between the Old and New Worlds. When he compared cultural features of the ancient Near East and Mesoamerica, he found a large number of significant parallels. Although Sorenson did not mention the fact, many of these features had in fact been raised and discussed by ES. They included: temple and temple platforms; swastika; astronomy, calendar and writing; astrological almanac; burial tombs and chambers; incense use in rituals; paradise and underworld concept; serpent and snake symbolism, dragon motif; double-headed eagle; winged sun disk; stele as cult objects; human and animal figurines; sacrifice complex (including blood offering, human and child sacrifice); libation vessels; dualism; kingship complex; purple dye; turban; weapons armour and trepanation.

Sorenson felt that his material was enough to challenge the (largely unexamined) orthodox view that any external influences upon Mesoamerica were minor or tangential, not seriously affecting indigenous tradition (Glyn Daniel, as we have seen, took this view). As Sorenson said, even if the style of Mesoamerica was highly distinctive, "some rather basic ideas seem to have been shared in the two areas". Unless further research disqualified his evidence "one must conclude that a substantial number of cultural features of much more than peripheral significance" in Mesoamerican civilization "either originated or were at least present even earlier" in Old World culture . He cited with approval Alfred L. Kroebner's dictum: "There is thus as much evidence needed for an assumption of independent origins as of a connection: the burden of proof is equal". Sorenson saw American anthropology as engaged in other projects. The evolutionary school saw cultural-historical data as useful in illustrating evolutionary theory. Social anthropologists were preoccupied with "clarifying local and regional cultural sequences. Neither has given serious, consistent consideration to the possibility of significant communication of culture from the Old to the New World".[17]

The hard-headed sceptics pointed to one big headache for diffusionists. This was the lack of verified archaeological finds of artifacts that might be expected from inter-hemisphere contacts. Ekholm, however, pointed to the incompleteness of the archaeological record, possible loss of artifacts over time and the non-material nature of many contacts. Vast areas had yet to be explored (for instance, in western Mexico and southeast Asia). From another perspective, it was difficult to deny the artistic similarities between Asia (especially China and Japan) and Mesoamerica: "scroll patterns from the Ulua Valley compared to Chou China, friezes from Tajín compared to Chou; pyrite mirrors, Tajín compared to China; tripod vessels, Teotihuacán style and Han China; features of temples, Yucatán Maya compared to Cambodia . . . Valdiva and Jomon (Japanese) ceramics" and much else.[18] Others suggested a link-up in boat types, axes and adzes, calendrical systems and pottery. There were some interesting linguistic similarities, or claimed similarities, between New and Old World. Much ink had been spilt on the resemblance between (say) the Quechua name for sweet potato, *kumar*, and the Polynesian name for it, *kumara*. Some scholars had gone to the edge of obsession trying to compare linguistic systems. However the feeling of the symposium was that the whole linguistic issue was as yet inconclusive. The same judgment was made about efforts by physical anthropologists to compare physical types across the hemispheres. We might add, from our present-day standpoint, that DNA research promises to solve that problem once and for all.

No one at the symposium held to the unicentric position of those like Elliot Smith on Egypt or Raglan on Sumeria, "with their underlying assumptions about the uninventiveness of man". (Defenders of ES might have pointed to the tendency to caricature ES's position. There was no systematic analysis of his more nuanced works, such as *Elephants and Ethnologists*.) However, as the editors noted, it was time to get past the old animosities between diffusionists and independent inventors: "The tendency to treat diffusion versus independent invention of culture traits or complexes may in fact obscure some of the problems of the interpretation of cultural interaction". The debate over pre-Columbian contacts had generated "more heat than light". It was time to move on to the larger question of cultural dynamics. The editors summarized the sense of the symposium with admirable lucidity. It is worth quoting at some length:

> The question of sporadic pre-Columbian contacts is interesting but ultimately not very important. Those members of the symposium who feel most strongly about the validity of transoceanic diffusion normally consider that contact was massive and prolonged, that it may have come in several waves, and that it profoundly influenced

> the art, architecture, political and social systems, crafts, religion, and other uses of culture . . . Those members less convinced of massive diffusion are particularly puzzled at the lack of hard (artifactual) evidence for contact, the curious dissimilarity of agricultural systems between the hemispheres, the increasing evidence of formative cultural sequences in the New World that match those already delineated in the Old World, and, of course, the real space and time problems of massive contact. . . .
>
> The underlying problem that must be solved is that of the origins of New World civilization. Clearly, the present status of our knowledge of American archaeology does not allow us to attribute the origins of New World civilization to diffusion from the Old World with assurance. Equally, however, it does not demonstrate the independent origin of New World high culture. Just as the zero occurrence of artifacts originating in the Old World and found in America may be taken as a strong argument against diffusionist explanation, so the early occurrence of a complex of Old World-like traits – often very sophisticated – in early levels of Nuclear American civilization casts a strong reflection against the independent origins hypothesis. Of course, it is also possible that the New World was a 'laboratory' situation as regards *origins* of civilization but received significant influence from the Old World at one or more later but still pre-Columbian dates.

What of future research? There was a call for two things: a seriously inter-disciplinary approach, using anthropology, botany, geography, history, paleontology and zoology; and a call for open minds: "we must particularly guard against ideas becoming so imbedded that they are accepted as gospel without check or challenge".[19] These were things that ES had always pleaded for.

There are signs that they are happening. In the American area, the old isolationist paradigm has been seriously undermined. A short *Afterword* is not the place to go into depth about research to the present. But one could cite, for example, the Harvard scholar David Kelley, whose meticulous research has shown irrefutable similarities between Eurasian and Mesoamerican calendar astronomy and astrology systems. (ES, following Zelia Nuttall, had long ago pointed to such parallels.) The parallels are so striking and intimately related that it defies belief that the systems could have been independently invented. Very interestingly, Kelley's considered view is that linkages can be traced between northern India and the area around Guatemala about two thousand years ago. One could cite the work of Terry Jones and Kathryn Klar on linguistic and archaeological evidence for early Polynesian contacts with Southern California.

Then there is Alice Kehoe's ground-breaking research on trans-oceanic voyaging in prehistoric America, a topic she has been engaged upon since the early 1960s.

I will content myself with giving a snapshot of the existing situation by reference to Alice Kehoe's refreshingly open-minded little book *Controversies in Archaeology* (2008).[20] A highly reputable scholar, Kehoe nevertheless encountered persistent resistance from mainstream academe for her unconventional ideas. She had two disadvantages: she was a woman in an academic world dominated by males (at least in the 1960s and 1970s); and she was proposing ideas on trans-oceanic movements of people and ideas that were outside the ruling paradigm. She describes how she was "a person not quite routinely accepted into American archaeology's core system . . . Experiencing unjust bias firsthand heightened my thinking critically about the status quo, whether about the people acclaimed as leaders (like the professor who would not accept women) or about core ideas". As a college student she worked as an aide in the anthropology department at the American Museum of Natural History, "where I observed a fine archaeologist and scholar, Gordon Ekholm, rebuffed when he put forward evidence for pre-Columbian trans-Pacific voyages. My undergraduate exposure to the scrupulous science of Ekholm, his colleagues James Ford and Junius Bird, and my professors Richard and Natalie Woodbury helped me to distinguish between sound work, sloppy thinking, and fads in archaeology" (p. 17).

What then is Kehoe's take on the controversy? She is an ardent apostle of the view that people crossed the oceans to America before Columbus. She bases this on decades of research. Her comments are sometimes uncannily reminiscent of ES.[21] For instance: "Common sense points to the indisputable fact that men and women crossed open ocean to get to Australia fifty thousand years ago in the Pleistocene. There is no other way that continent could have been populated. Polynesians sailed to hundreds of islands in the Pacific many centuries before Europe's Age of Exploration began in the fifteenth century" (p. 140). She says a little later: "if for centuries, Polynesians deliberately explored the entire Pacific Ocean and had the technology to navigate precisely and transport settlers with their tools and foodstuffs, does it seem likely that in all those centuries, not one Polynesian ever discovered America?" (p. 141).

Her overall position may be encapsulated thus:

> Our own experience and common sense tell us how commonplace it is to get a new idea or technology from someone in another culture, whether it's pizza and tacos, or yoga, or a Japanese video game. Conversely, it is rare to independently invent something – if you try to get a patent on an invention, you'll find it isn't easy to

> prove originality. Nevertheless, it has been conventional in American archaeology to assert that the Americas were isolated from contacts with other continents after initial population migration over the Bering Strait into Alaska around ten thousand ago, and that consequently, all the native societies of the America independently invented everything in their cultures other than the Ice Age hunters' equipment. [In her book she looks at boat technology] and at evidence, some beyond doubt and much controversial, for intersocietal contacts before the European Age of Exploration and colonization. While many of my colleagues don't agree with all of the claims I make . . . , I believe that the model of intermittent contacts best explains the cited features of indigenous American cultures. (p. 140)

There is now voluminous data on pre-Columbian transoceanic contacts (she cites the work of the historical geographer Stephen C. Jett in this field). Her own evidence covers such examples as the unusual sewn-plank-built canoes shared by the Polynesians and the Chumash and Kumivit of southern California (the Californian names for the canoes are very similar to the Hawaiian names for the same thing); and the sweet potato, which is native to South America and spread though the Pacific about a thousand years ago: "There is no way sweet potatoes spread through the Pacific except by being carried in boats, and sharing the same word for the plant [*kumara*] implies peaceful trading contact" (p. 142). There is supporting evidence also from plant and animal organisms.[22]

In an interesting theoretical move, Kehoe makes use of a concept developed by Joseph Needham, the famous historian of Chinese science. Needham studied the diffusion (and the non-diffusion) of Chinese science and technology to the rest of the world up until the eighteenth century. He and his co-worker Lu Gwei-Djen devised tests to try to establish whether (say) an invention (one could also apply this to a cultural feature) had in fact been diffused from another source: "One criterion is *collocation*, the number of items or traits together (collated) in the feature; the more complex the artifact or steps in its manufacture, or the larger the number of items that occur together, the higher the probability of transmission together" (p. 146). ES would have welcomed this idea with open arms. His concept of cultural clusters (as visualised on his maps depicting the spread of groups of Egyptian traits across the world) was in fact a groping towards such a concept as collocation.

Kehoe postulates a number of complexes that agree well with the principle of collocation. Together they make (she says) a compelling case for intersocietal contacts across major bodies of water. They include papermaking, royal purple dye and wheeled figurines (see pp. 159–160 for her evidence). More examples include fighting cocks fitted with Asian-style spurs and distilling liquor through Asian-type pot stills. She comments on other links:

> A number of customs in Mesoamerica are particularly Asian-like, including valuing jade highly and placing a jade bead painted red with cinnabar in the mouth of corpses; building tiered pyramids symbolizing the seven or nine or thirteen heavens; formal body and hand positions (called *mudra* in India) seen in Mesoamerican art, and the ruler seated on his throne with one leg tucked under, the other hanging. Other customs, like calendar astrology . . . are widespread in Eurasia, including the Tree of Life with a great bird on top, a lion or jaguar on earth, and a serpent among its roots. Still others are more contested, including the identification of ears of maize (corn) carved on medieval temples in South India, and peanuts – a South American crop – recovered from archaeological excavations in China. Peanuts and other American crops continue to turn up in Eurasian and Oceanic excavations, substantiating references in pre-Columbian Chinese and Indian texts on cultivated plants. Taken together, there is strong evidence for multiple transoceanic contacts and borrowings before 1492, and, at the same time, absolute evidence that American civilizations developed independently of any attempts at colonization by Eurasian or African nations. What the evidence shows is that America's indigenous nations were part of global connections for several thousand years before Columbus kicked off the historic invasions. (p. 160)

And so the debate continues.

In conclusion I would simply say that Grafton Elliot Smith's reputation deserves rehabilitation. Admittedly, ideas that he raised in his now-forgotten books keep resurfacing in the recent literature, but this is rarely acknowledged. The grotesque caricatures and stereotypes of him need to end. Elliot Smith was a serious scientist in all the fields he tackled. It is grossly unfair to excise his ethnological work as some sort of aberration. In this field he devoted enormous energy to collecting as much reliable data as was then available. And he applied his formidable intelligence to it. He put forward innovative hypotheses based on such observed evidence. As a scientist he was fully aware that such hypotheses would survive only until they were disproven (scientific laws cannot be proven, only disproven). If he sometimes speculated beyond his data, he did so with the purpose of stimulating debate and more intensive research. Only by means of such dynamic dialogue and rigorous testing would a truer understanding of human history emerge. As we have seen, much of his work was appreciated at the time, even by later critics such as Malinowski. But he lived to see his paradigms overtaken by new paradigms. They were concerned with essentially different things, and asked different questions. He fought this, with all his characteristic vigour but ultimately unsuccessfully. He complained long and hard that emerging academic cultures increasingly shunned outsiders like him with unfashionable ideas. He had a wider, more inclusive view of knowledge. While his sweeping Egyptology theory will never resurface, many of his observations, examples and speculations continue to offer signals for future research.

Notes

1 Early Days: Sydney, Cairo, Manchester (1871–1915)

1 Biographical material comes from the *Dictionary of National Biography, 1931–1940*, the *Oxford Dictionary of National Biography* (2004), obituaries and various other sources. Major biographical information is contained in Warren R. Dawson, ed., *Sir Grafton Elliot Smith: A Biographical Record By His Colleagues* (London: Jonathan Cape, 1938) [hereafter cited as Dawson]. Warren Dawson was a student and colleague of ES's and had access to much of ES's correspondence and other materials, some of which are now unavailable. His long chapter I, "A General Biography", pp. 17–112, is a valuable source. There is also material in other commemorative volumes: Professor Lord Zuckerman, ed, *The Concepts of Human Evolution* (London: Academic Press, 1973), proceedings of a symposium in honour of ES organized by the British Anatomical Society and the Zoolological Society of London, held in 1972; and A. P. Elkin and N. W. G. Mackintosh, eds, *Grafton Elliot Smith: The Man and His Work* (Sydney: Sydney University Press, 1974),especially the essays by Elkin and Raymond Dart, both taught by ES.

2 In his marriage certificate of 2 September 1900, Scottish National Church, Chelsea, London, his surname is given as "Elliot Smith": Marriage Certificate 92318/COL651652, *England and Wales,Marriage Index, 1837–1915*. His wife's name is given as Kate Emily Macredie (rather than the Kathleen usually cited).

3 New South Wales Death Certificate, Reg. No 016582. He died in Sydney on 27 July 1929, aged 89 years. His parents were Elliot Smith, confectioner, and Eliza née Sheldrick.

4 ES, "Fragments of Autobiography" in Dawson, pp. 113–120 (quote p. 113).

5 New South Wales Death Certificate, Reg. No. 000362. He died in Sydney on 26 February 1880, aged 63. His parents were James Smith, gardener, and Rebecca née Lofts. He was born in Cambridgeshire, England, and lived 23 years in Australia. His occupation then was listed as baker and pastrycook.

6 The family moved from Melbourne to Sydney in December 1857 on the 393 ton steamship *Wonga Wonga*. The children's names were given as Mary Ann, Susan, Caroline and S. S. Smith: *New South Wales Unassisted Immigrant Passenger Lists, 1826–1922*.

7 ES, "Fragments", p. 114.

8 New South Wales Marriage Certificate: Reg. No. 000608. They were married on 20 July 1864 at the Pitt Street Congregational Church, Sydney (although he was residing at Wollombi, New South Wales). He apparently remained a Congregationalist as his death certificate was signed by a Congregational minister.
9 ES also had two elder sisters Maud and Caroline and a younger sister Lily (Stephen Sheldrick Smith death certificate).
10 ES, "Fragments", p. 114.
11 ES, "Fragments", p. 115.
12 ES, "Fragments", pp. 117–118; J. T. Wilson, "Sir Grafton Elliot Smith", *Obituary Notices of Fellows of the Royal Society*, 2 (January 1938), pp. 323–324.
13 Wilson, "Sir Grafton Elliot Smith", p. 324.
14 See James Belich, *Replenishing the Earth: The Settler Revolution of the Anglo-world* (Oxford: Oxford University Press, 2009).
15 Dawson, p. 18.
16 H. D. Black (Chancellor of Sydney University), "Introduction", Elkin and Mackintosh, eds, *Elliot Smith*, p. 5.
17 ES to Robert Broom, (18 August 1896), quoted Dawson, p. 20.
18 ES to J. T. Wilson (7 September 1896), quoted Dawson, p. 21.
19 J. T. Wilson, "Sir Grafton Elliot Smith", p. 325.
20 Lord Rutherford, "Early Days in Cambridge" in Dawson, pp. 134–135. Rutherford died a few days after penning this short memoir for Dawson's book.
21 Dawson, pp. 24–25. He gave a course of lectures on the central nervous system, and also did research in the museum of the Royal College of Surgeons.
22 ES to J. T. Wilson (5 February 1899), quoted Dawson, p. 27.
23 Matthew Young, "Obituary" in *Man*, 37 (March 1937), p. 51.
24 Dawson, p. 28.
25 Young, "Obituary", p. 51. (Note that this was written in 1937).
26 Macalister to ES (4 July 1900), quoted Dawson, p. 29 (and generally for events at this time).
27 ES to J. T. Wilson (24 February 1901), quoted Dawson, p. 32.
28 ES, "Introduction", W. H. R. Rivers, *Psychology and Ethnology* (London: Kegan Paul, 1926), a collection of essays that ES was instrumental in having published as a book after Rivers's death, pp. xiii-xiv. ES makes the point about Rivers's calling his attention to the preservation of the brain in ancient Egyptians in his introduction to Rivers's *Psychology and Politics and other Essays* (London: Kegan Paul, 1923), also edited by ES, p. 126, note. ES said that Rivers had "happened to see dessicated brains in the skulls brought to light there [Upper Nile] by Dr Randall-Maciver". See essay 6 for Rivers's account of ES's Egyptian work and diffusionism (pp. 126–129). It has been suggested that Rivers also urged ES to go up the Nile to look at mummies at Flinders Petrie's digs: Richard Slobodin, *W. H. R. Rivers* (New York: Columbia University Press, 1978), p. 148. Glyn Daniel (a strong critic of ES's theories) denied that ES ever visited any of Petrie's excavations, and claims that the two men were at loggerheads

throughout their lives: "They clashed over the nature and techniques of Egyptian mummification, but most of all over ES's extravagant theories of the Egyptian origins of all things": *Nature*, 317 (5 September 1985), p. 26.

29 For instance, "On the Natural Preservation of the Brain in The Ancient Egyptians", *Journal of Anatomy and Physiology*, 36 (1902), pp. 375–380.
30 Ian Langham, *The Building of British Anthropology* (London: Reidel, 1981), p. 136.
31 Dawson, pp. 38–39. ES gave more details in a letter to the *Times*: "The incident took place in Cairo in 1903, when the mummy of the Pharaoh Thothmes IV was taken by Mr. Howard Carter and me to the late Dr. Herbert Milton's private nursing home, where an excellent radiologist, Dr. Khayat Bey, was in charge of the only X-ray apparatus then in Egypt. The photographs were taken for the purpose of determining the Pharaoh's age at the time of death, a question concerning which the historians were in doubt": *Times* (6 March 1935), p. 15. The *Illustrated News* published some of the photographs in 1923.
32 J. T. Wilson, "Sir Grafton Elliot Smith", pp. 327–328.
33 As an example we find him writing: "There were 64 huge cases of prehistoric remains from Upper Egypt to be unpacked and arranged in Museum shape; there was a bundle of 38 drawings for a first Sphenodon paper to be redrawn for Howes: and worse than all, I had to conduct an examination and start a new session with an unprecedently large enrolment of new students. For the past five weeks I had literally not a moment to call my own": ES to Robert Broom (5 November 1902), quoted Dawson, p. 36. Further ES letters cited in the text are from correspondence collected in Dawson's biographical sketch. As readers can readily consult them there (the sketch is chronologically ordered), I will henceforth simply cite them in the text by date and recipient.
34 Wood Jones detailed his collaboration with ES in the Nubian Survey in his chapter, "In Egypt and Nubia" in Dawson, pp. 139–150.
35 Wood Jones, "In Egypt and Nubia", pp. 139, 147–148.
36 For details of his administration see the chapter by J. S. B. Stopford (later Vice-Chancellor and professor of anatomy at Manchester), "The Manchester Period" in Dawson, pp. 151–168.
37 T. H. Pear, "Some Early Relations between English Ethnologists and Psychologists", *Journal of Royal Anthropological Institute of Great Britain and Ireland*, 90 (1960), pp. 228–229. The ethnologists W. J. Perry and C. G. Seligman seem to have been on the fringes of the group (pp. 229–230).
38 Rivers published a paper "The Disappearance of Useful Arts" in 1912 and one on "The Contact of Peoples" in 1913: see "The Aims of Ethnology" in his *Psychology and Politics* (Cambridge: Kegan Paul, 1923), pp. 114, 117.
39 Pear, "Some Early Relations", p. 232.
40 ES, preface to *The Ancient Egyptians and the Origin of Civilization*, 2nd ed., (London, New York: Harper, 1923), pp. v–vi.
41 W. J. Perry, "Anthropologist and Ethnologist" in Dawson, p. 206.

42 See, for example, ES, "The Evolution of the Rock-cut Tomb and the Dolmen", in *Essays and Studies presented to William Ridgway* (Cambridge, 1913), p. 493.
43 Lewis gave another paper to the British Association in 1911 which reviewed a large number of dolmens and other stone monuments. Because of the many local differences in construction and, apparently, purpose, and the vast range over which they ranged, he concluded: "... it is probable that they were not the work of a single race, which went about the world constructing them ... but that they were part of a phase of culture through which many races have passed. Little if anything can be deduced from these monuments as to early migrations of the human race": A. L. Lewis, "Dolmens or Cromlechs", *Man*, 11, 98 (1911), p. 175. ES returned to the fray at the British Association in 1913, comparing Egyptian with Sardinian tombs: "The Origin of Dolmens", *Man*, 13 (1913), pp. 193–194.
44 ES, preface to his *Ancient Egyptians* 2nd ed, 1923, pp. vi–vii.
45 W. H. R. Rivers, "The Ethnological Analysis of Culture", *Journal of British Association for the Advancement of Science* (1911), pp. 490–499; also Rivers, *Psychology and Politics*, pp. 117–118.
46 ES, preface to *Ancient Egyptians* , 2nd ed., 1923, pp. vii–viii.
47 *Ibid.*, p. ix. See also W. H. R. Rivers, *Psychology and Politics* (London: Kegan Paul, 1923), p. 116.
48 In this he followed Reisner's conclusion, based on excavations at Naga-el-Der, that the Upper Nile was the origin of copper-working. Not all experts accepted this. See J. L. Myres's review of *Ancient Egyptians* in *Man*, 12 (1912), p. 187. Myres praised the book as an impressive synthesis by a non-expert, but claimed that ES underestimated the antiquity of copper implements in Babylonia. On this and other arguments ES "conveys the impression that the basis of observed fact is still precarious": pp. 188–189. See also review by H. R. Hill in *Nature*, 88 (8 February 1912), p. 475, which described the book as "one of the most important recent contributions to Egyptian archaeology". Recent research has suggested the possibility that copper mining and working may have existed as early as the fifth millennium BC in the Danube valley: Daniel W. Anthony with Jennifer Chu, eds, *The Lost World of Old Europe: The Danube Valley, 5000–3500* BC (Princeton University Press, 2010).
49 Elliot Smith, "The Foreign Relations and Influence of the Egyptians under the Ancient Empire" (British Association meeting, 1911), *Man*, 11 (1911), p. 176.
50 ES, *The Ancient Egyptians and Their Influence upon the Civilization of Europe* (London, New York: Harper, 1911), p. 59. (He gave a more ambitious title to the second edition of 1923: *The Ancient Egyptians and the Origin of Civilization*. This second edition also added a last chapter on the tomb of Tutankhamen).
51 Dawson, p. 54.
52 Graeme L. Pretty, "The Macleay Museum Mummy from Torres Straits", *Man*, ns 4 (March 1969), pp. 27, 39. His colleague T. H. Pear agreed that ES's views might have been more warmly received if he had been less

pugnacious. He described ES as "a bonhomous good mixer who loved a fight . . . Though in the actual presence of people his urbanity was as marked as a Foreign Office representative's, he was no respecter of persons if the criterion of their value was intellectual": Pear, "Some Early Relations", p. 228. Ian Langham rather belligerently described ES as "a bombastic controversialist". Langham blamed ES for leading Rivers astray from his more rigorous genealogical method: *The Building of British Social Anthropology*, p. 142 and *passim*.

53 Zuckerman, ed., *The Concepts of Human Evolution*, pp. 6–7.

54 Dawson, p. 55.

2 Migrations of Early Culture (1915)

1 Grafton Elliot Smith, *The Migrations of Early Culture: A Study of the Significance of the Geographical Distribution of the Practice of Mummification as Evidence of the Migrations of Peoples and the Spread of certain Customs and Beliefs* (Manchester: Manchester University Press; London, New York: Longmans, Green, 1915). This was originally a paper read to the Manchester Literary and Philosophical Society on 23 February 1915, published as a paper in *Memoirs and Proceedings* of the society, Vol. 59, no. 10 (1915), pp. 1–143. The book was reprinted from this paper by Manchester University Press using the same pagination. Page references in brackets in the text following refer to this book.

2 "In 1851 the German ethnologist Adolf Bastian (1826–1905) began a series of voyages around the world in order to build up the collections of the Royal Museum of Ethnology in Berlin. Impressed by the cultural similarities that he encountered in widely separated regions, he emphasized the Enlightenment doctrine of psychic unity by arguing that as a result of universally shared 'elementary ideas' (*Elementargedanke*) peoples at the same level of development who are facing similar problems will, within the constraints imposed by their environments, tend to develop similar solutions to them": Bruce G. Trigger, *A History of Archaeological Thought* (Cambridge University Press, 1989, repr. 2006), pp. 99–100.

3 C. L. Riley *et al.*, eds, *Man Across The Sea* (Austin, London: University of Texas Press, 1971), p. xii. The English ethnologist Edward Tyler had also put forward similar ideas, although his ideas on diffusion were regarded by ES as ambiguous (discussed in a later chapter): "According to Tyler, the probability of [cultural] contact increases in ratio to the number of arbitrary, similar elements in any two trait complexes" (*ibid.*, p. xii).

4 "Does any theory of evolution help in explaining these associations? They are clearly fortuitous associations of customs and beliefs, which have no inherent relationship one to the other. They became connected purely by chance in one definite locality, and the fact that such incongruous customs reappear in association in distant parts of the globe is proof of the most positive kind that the wanderings of peoples much have brought this peculiar combination of freakish practices from the centre where chance linked them together" (pp. 6–7). Reinforcing this was the fact that in many cultures there were oral traditions that centred on

"culture-heroes" who were said to have brought in such customs and at the same time introduced knowledge of agriculture and weaving (which at the time were generally agreed to have originated with the Egyptians).

5 W. H. R. Rivers, *Psychology and Politics* (London: Kegan Paul, 1923), pp. 122–123. These remarks were made in a lecture Rivers wrote in 1919. He added that religious motives probably played a greater role in ancient times. Even gold was sought for its supposedly magical or religious qualities, while a possible motive was the widespread search for "an elixir of life". Spices and odorous resins were needed for preserving the dead. Perry published his views in a paper "The Relationship between the Geographical Distribution of Megalithic Monuments and Ancient Mines", *Memoirs and Proceedings of the Manchester Literary and Philosophical Society* (1915).

6 W. H. R. Rivers later commented that ES's findings showed "that the practices used by the people of Torres Strait in order to preserve the dead agree with those of Egypt in no less than seven points of detail. If we accept the independent origin of mummification in Torres Strait, we are forced to believe that, in a climate most unsuited for such experiments, the rude savages of those islands invented a procedure which took the highly civilized Egyptian many centuries of patient research to attain": *Psychology and Ethnology* (London: Kegan Paul, 1926) [ed. and with preface and introduction by ES], pp. 168–169. See also Ian Langham, *The Building of British Anthropology: W. H. R. Rivers and his Cambridge Disciples in the Development of Kinship Studies, 1898–1931* (London: Reidel, 1981), pp. 139–141. Dawson later examined a similar mummy that was held in The British Museum. See Dawson, "A Mummy from the Torres Straits", *Annals of Archaeology and Anthropology*, 11 (1924), pp. 87–96.

7 Letter from D. M. S. Watson to Warren Dawson (26 September 1937), cited in Dawson, pp. 63–65.

8 See Graeme L. Pretty, "The Macleay Museum Mummy from Torres Straits: A Postscript to Elliot Smith and the Diffusion Controversy", *Man*, ns 4, 1 (March 1969), pp. 24–52; also C. E. Joel, "Elliot Smith and the Macleay Mummy", *Man*, ns 4, 4 (December 1969), pp 645–46; G. Kraus, *ibid.*, pp. 646–47; A. T. Sandison, *Man*, ns 5, 2 (June 1970), p. 309; C. E. Joel and A. T. Sandison, *Man*, ns 5, 4 (December 1970), pp. 703–04. Pretty did a detailed reexamination of the Macleay mummy in 1962–63. Although his findings did not differ materially from those of ES, he suggested that ES may have fundamentally misinterpreted the Papuan artifact, underestimating the indigenous capacity to improvise, using imagination and intelligence (p. 39). Joel and Kraus argued that Pretty's findings did not essentially conflict with ES's position. Kraus felt that comparison "leaves little doubt in one's mind about the Egyptian derivation of the techniques employed in the embalming of the Macleay and other mummies from Torres Straits" (p. 648). Pretty concluded that ES's "diffusion heresy" still had contemporary relevance, as anthropology moved away from evolutionary stage theory that ranged societies in an order of linear progression: "Cultural links across wide distances separated by marked varieties in racial type make discussion of the Pacific

Ocean and its fringes a continuing source for theoretical controversies. [The whole viewpoint of anthropology] now appears more as a comprehensive and constantly fragmenting querying about man and his activities in all his variety, to see him in his own terms and ourselves in his. Anthropologists are returning to history as an essential background to the study of societies in a state of change . . . which makes study of Elliot Smith's ideas worth our notice" (pp. 39–40).

9 He presumably added his comments on the mummy after giving his original paper: "The Ancient Inhabitants of Egypt and the Sudan", *Man*, 14 (1914), p. 172.

10 Perry published his thoughts in *The Megalithic Culture of Indonesia* (Manchester, 1918). Rivers gave a pivotal role to Perry in his *Psychology and Politics* (1923), pp. 118–124. It had been thought that any migration of megalithic art must have come to the Pacific via Japan, given the supposed absence of megalithic handiwork in the Indonesia area. Perry had shown that there were small scale dolmens resembling Egyptian stonework in Indonesia, such as stone seats and offering places. He explained the modesty of such stonework in terms of local conditions (p. 120).

11 Clayton Joel, "William James Perry, 1887–1949", *Man* (January 1950), pp. 6–7; Daryll Forde, "Dr. W. J. Perry", *Nature*, 163 (4 June 1949), pp. 865–866; *Nature*, 111 (31 March 1923), p. 449. Joel commented: "Perry's was a happy and friendly character, blended of humour, forthrightness and keen perception, and impatient of cant and humbug. Despite the violence of the controversies in which he was involved over the Diffusion issue, he maintained the most friendly relations with most of those who publicly disputed his views". Illness compelled him to retire in 1939. His papers are at University College London.

12 *Nature*, 96 (11 November 1915), p. 300. Rivers's comments are reprinted in full as essay 4 of his *Psychology and Ethnology* (1926), "The Distribution of Megalithic Civilization". He explained that ever since he became interested in the blending of peoples as a dominant factor in human progress he expected megalithic monuments to provide "convincing demonstrations" of this. For several years he had been convinced "that the ideas underlying the construction of megalithic monuments had their origin in some one part of the world, whence they spread to other parts". But he hesitated to follow ES in his conclusion that Egypt was the home of megalithic culture. He was impressed by ES's dolmen theory "but it did not seem to me sufficient to prove conclusively that the mastaba was the prototype and original ancestor of the world-wide dolmen". He explained his hesitation as a reaction against "the many wild and uncritical attempts which have long been made to derive practices of remote parts of the world from Egypt". He also had "so keen a sense of the complexity of human progress" that he distrusted ES's simple explanation. Even after he accepted ES's theory that the dolmen and mastaba were "genetically related", his first inclination was "to regard the mastaba as the result of a special line of development in ideas which elsewhere produced the dolmen, and I preferred to remain in suspense concerning

the original home of these ideas rather than accept their origin in Egypt". As he explained to the 1915 BA meeting, he remained in this state of mind until ES spectacularly showed the resemblances between Egyptian and Papuan mummification in Melbourne in 1914, bolstered by his arguments in *Migrations of Early Culture* (Rivers, pp. 167–168).

13 Margaret S. Drower, *Flinders Petrie: A Life in Archaeology* (Madison, London: University of Wisconsin Press, 2nd ed., 1995; 1st ed., 1985), p. 347. The antagonism between ES and Petrie began in 1910 when ES declared as "nonsense" Petrie's claim that finds at Meydum and Mastaba supported his theory of ritual dismemberment. ES had never seen a single case of such practice after examining thousands of mummies. A heated exchange in letters and the journals ensued. General opinion has been against Petrie's theory. The two men were colleagues at University College and for many years were barely on speaking terms (pp. 345–347). According to Glyn Daniel, Petrie's copy of the 1923 edition of ES's *Ancient Egyptians* "is peppered with outspoken comments such as, 'No such thing', 'Nonsense', 'What a romance', 'no evidence', 'No, No' . . . ": Glyn Daniel, "ES, Egypt and Diffusionism" in Solly Zuckerman, ed, *The Concepts of Human Evolution* (London: Academic Press, 1973), p. 417.

14 *Nature* (30 September 1915), p. 131. Seligman and Rivers had been members of the famous Torres Straits expedition of 1898, together with other prominent figures such as A. C. Haddon, C. S. Myres and William McDougall. For the papers given see ES, "Discussion on the Influence of Ancient Egyptian Civilisation on the World's Culture", *Report of the 85th Meeting of the British Association for the Advancement of Science (Manchester 1915)* [London: Murray, 1916], pp. 667–669; and W. J. Perry, "On the Influence of Egyptian Civilisation upon the World's Culture", *ibid.*, pp. 669–670.

3 Dragons and Critics (1915–1920)

1 He was much struck by "her great monograph": ES, "The Influence of Ancient Egyptian Civilization in the East and In America", *Bulletin of John Rylands Library*, 3 (1916–1917), p. 61; Zelia Nuttall, "The Fundamental Principles of Old and New World Civilizations: A Comparative Research based on a Study of the Ancient Mexican Religious, Sociological, and Calendrical Systems", *Archeological and Ethnological Papers of the Peabody Museum, Harvard University*, 2 (March 1901).

2 ES, *The Evolution of the Dragon* (ManchesterManchester University Press, 1919), p. xiii.

3 ES, "The Influence of Ancient Egyptian Civilization in the East and America", pp. 48–77 (quote p. 73 n1), published by Manchester University Press. Page references in brackets in the text below are from this source.

4 ES took much material on the Phoenicians from Nuttall, without accepting some of her more extreme astronomical speculations. He also used the work of scholars such as Hogarth, Siret, Dahse and Sayce. There

had been a revisionist move within classical scholarship to downplay some exaggerated claims that had been made for the Phoenicians. ES accepted the view that more credit needed to be given to the Egyptians, Minoans and other mariners without detracting from the real exploits of the Phoenicians.

5 ES was influenced here by the work of Alan H. Gardiner. He had given a paper at the British Association meeting at Manchester in 1915 in which he argued (on the basis of scripts recently discovered in Sinai) that Phoenician, Greek and Minoan letters were borrowed from the Egyptian hieroglyphic system.

6 A custom in some indigenous cultures whereby, on the birth of a child, the father is put to bed and treated as if he were physically affected by the birth. The English ethnologist E. B. Tyler coined the name for this "man-childbed" (OED).

7 ES, "The Origin of the pre-columbian Civilization of America", *Science*, ns 44 (11 August 1916), pp. 190–195.

8 A. A. Goldenweiser, "Diffusion versus Independent Origin: A Rejoinder to Professor G. Elliot Smith", *Science*, ns 44 (13 October 1916), pp. 531–533. (ES later named Flinders Petrie and Cecil Firth as coming close to claiming that humans had an instinct to build chambered tumuli.)

9 In a review of Rivers's monumental *History of Melanesian Society* (1914), Goldenweiser mixed praise with criticism of Rivers's "reckless" use of diffusionist method (in volume two): "Deliberately evading any attempt to furnish proof of diffusion in specific instances, the author erects a purely hypothetical structure, based on a bewildering maze of assumptions invariably favoring interpretations through diffusion while disregarding alternative interpretations": *Science*, ns 44 (8 December 1916), pp. 824–828 (quote p. 827).

10 Philip Ainsworth Means, "Some Objections to Mr. Elliot Smith's Theory", *Science*, ns 44 (13 October 1916), pp. 533–534.

11 ES, "The Origin of the pre-columbian Civilization of America", *Science*, ns 45 (9 March 1917), pp. 241–246.

12 Based on his lecture to the Manchester Literary and Philosophical Society at the Rylands Library on 9 February 1916.

13 ES, *The Evolution of the Dragon* (Manchester, Manchester University Press; London, New York: Longmans, Green, 1919), pp. 71–72 (quotes), 10. Page references in the text are to this source.

14 The words are those of W. R. Lethaby, quoted p. 12. Lethaby did not suggest a mummification link, but ES was impressed by his claim that the origins of architecture were to be found in Egypt. Lethaby traced the square-roomed house, columnar orders and fine masonry to Egypt. He said "the mission of Greece was rather to settle down to a task of gathering, interpreting, and bringing to perfection Egypt's gifts" (quoted p. 11).

15 This chapter was an elaboration of a lecture he gave in the John Rylands Library in Manchester on 8 November 1916.

16 H. J. Spinden, "The Prosaic vs. the Romantic School in Anthropology"

in ES, ed., *Culture: The Diffusionist Controversy* (New York: Norton, 1927), p. 60.

17 It was based on his lecture at the John Rylands Library on 14 November 1917.

18 J. Wilfrid Jackson, *Shells as Evidence of the Migration of Early Culture* (Manchester: Manchester University Press, 1917), 216 pages. This was reprinted, with additions, from the *Proceedings of the Manchester Literary and Philosophical Society*. The title, of course, echoes ES's 1915 paper. Jackson (1880–1978) was a massive figure in conchology who did pioneering field work in Lancashire and Derbyshire. He was "an all-round naturalist in the best traditions of the late Nineteenth Century. To conchology he added expert knowledge of local natural history, cave Mammalia, archaeology and Carboniferous paleontology": "Obituary", *Journal of Conchology*, Vol. 30, p. 222. The Oxford anthropologist A. C. Haddon reviewed Jackson's book favourably, noting that Jackson worked along the same lines as ES's "very active school" of diffusionists at Manchester who claimed the primacy of Egypt. Haddon was particularly impressed by Jackson's evidence on purple dyes: "The employment of the pigment found in certain marine shells for dyeing fabrics was known in the Mediterranean area and West Britain, was practised in prehistoric Japan and still is in China, and also by pre-Columbian Incas, and in Central America". Haddon felt that while objects might be carried by a cultural drift, a special technique (such as dyeing, but also the moon-god cult) was probably transmitted by personal knowledge: "The cumulative evidence of ethno-conchology is too great to be ignored, and affords additional demonstration of the spread of a complex culture from the culture centres of the Old World to South and Central America": A. C. Haddon, review of Jackson's book, *Nature*, 100 (21 February 1918), p. 482.

19 M. J. Bishop, ed., *The Cave Hunters: Biographical Sketches of the Lives of Sir William Boyd Dawkins (1836–1929) and Dr. J. Wilfrid Jackson(1880–1978)* (Derbyshire Museum Service, 1982), p. 29: "Elliot Smith was interested in any areas of archaeology and ethnography that might support his diffusionist theories, and upon learning something of the cultural uses of shells in various corners of the world, approached Jackson to investigate the subject from a scientific and ethnographical point of view".

20 See Henrika Kuklick, *The Savage Within: The Social History of British Anthropology, 1885–1945* (Cambridge: Cambridge University Press, 1991), p. 162. As she points out, ES defended the psychoanalytical method for treatment of mental disorder, especially in his work on shell-shock during the Great War. But he dissented from Freud's emphasis on sex, insisting (she says) that fear was the only universal human drive. However "In fact, Elliot Smith's explanation of the appeal of certain cultural practices to peoples all over the world was evidently Freudian in content if not in name, depending as it did on the premise that peoples everywhere found compelling such natural symbols as the cowrie shell, suggestive of female genitalia". On the question of fear as a predominant drive, ES's position was more complex than Kuklick suggests. It is true

that he saw fear of death as a major explanation of ritual and religion. But overall he preferred to emphasise the overriding importance in early human history of the "Life Quest", and the image of deities as "Givers of Life". The instinct of sex was "imperious", but "it does not play the same kind of role as the more fundamental and continuously active instinct of self-preservation does": ES, *Human History* (London, 1930), pp. 33–34.

21 The eminent archaeologist W. M. F. Petrie questioned the pre-eminence of the cowry shell in prehistoric ideas. This he said was not reflected in the discoveries "where that shell is much rarer than others down to 900 BC, and the oldest gold cowry known is of the XIIth dynasty": review of *Evolution of the Dragon* in *Man*, 20 (May 1920), p. 76.

22 Gerhard Kraus, *Homo Sapiens in Decline* (Great Gransden: New Diffusionist Press, 1973), p. 144.

23 He also made the interesting observation that dragons (especially in China) were conventionally represented as red. This went back to the early practices of blood substitution in sacrifice (using red-stained beer, red wine, red earth, red berries). These life-giving and death-dealing substances were associated with red, and the destructive demons Sekhet and Set were given red forms. This in turn was transmitted to the dragon "and to that specialized form of the dragon which has become the conventional way of representing Satan" (p. 205). While his book was in proof ES came across a Mayan pot in the form of a crocodile-dragon in the Liverpool Free Public Museum. This had many features seeming to bear out the thesis of his book: see his "An American Dragon", *Man*, 18 (November 1918), pp. 161–166.

24 *Athenaeum* (27 June 1919), pp. 522–23. The reviewer accused ES of being in a hurry to jump to conclusions: "When Professor ES writes 'it is certain', he appears to mean 'I have just read something in a book making it not unlikely' . . . Though we cannot but admire the giant energy that thus piles Pelion on Ossa, we must protest in the name of all the decencies that the kingdom of Heaven suffereth violence". See also W. M. F. Petrie in *Man*, 20 (May 1920), pp. 76–77; and *Nature*, 104 (4 December 1919), pp. 350–51.

25 Alexander A. Goldenweiser, *Early Civilization: An Introduction to Anthropology* (London, *c.*1922), p. 311. The edition is undated but the preface is signed December 1921.

4 Elephants and Ethnologists (1920–1924)

1 "Obituary", *Nature* (9 January 1937), p. 58. I have drawn on this for other biographical information in the text.

2 "That Elliot Smith, probably the greatest neuroanatomist alive, should subject the Piltdown skull to thorough scrutiny and pronounce it ape-like is truly astonishing. The skull was merely that of a modern human being. The jaw, on the other hand, was not merely ape-*like:* It was *ape* (an orangutan, to be precise), carefully stained to look old, its teeth filed down to resemble the flat pattern of wear seen in human teeth. The bones had been deliberately placed together in the gravel pit by human hands,

so that they could later be 'discovered' by a credulous world. Not just Smith Woodward, Elliot Smith, and [Arthur] Keith were fooled by the deception, but the entire British scientific establishment as well": Donald Johanson and James Shreeve, *Lucy's Child: The Discovery of a Human Ancestor* (Bungay, Suffolk: Viking, 1990), pp. 51–52.

3 S. Zuckerman, "Sir Grafton Elliot Smith" in Zuckerman, ed., *The Concepts of Human Evolution* (London: Academic Press, 1973), p. 3.

4 "The Rockefeller Foundation Gift of the Institute of Anatomy to University College, London", *Nature* (2 June 1923), p. 755.

5 ES, "Medical Research", *Nature* (15 July 1920), pp. 605–607. See also Donald Fisher, "The Rockefeller Foundation and the Development of Scientific Medicine in Great Britain", *Minerva*, 16 (1978), pp. 20–41.

6 *Nature* (2 June 1923), p. 756 (possibly by ES).

7 *Nature* (31 March 1923), p. 449. ES's successor as professor of anatomy at University College, H. A. Harris, recalled that the establishment of a readership in anthropology had "raised a protest in many quarters". He added: "There is little doubt but that Elliot Smith intended to import W. H. R. Rivers from St. John's College, Cambridge, in order that he should continue his work on the psychology of primitive people." However Rivers "would not leave St. John's", so Perry was appointed: Dawson, p. 176.

8 "Sir Grafton Elliot Smith", *Oxford Dictionary of National Biography* (17 August 2006), p. 2.

9 Malinowski became lecturer in social anthropology at the London School of Economics in 1923, and foundation professor of social anthropology in 1927.

10 *Encyclopaedia Brittannica*, 12th edition (1920), pp. 147–148.

11 See A. C. Haddon, "Obituary of Sir E. B. Tylor (1832–1917)", *Nature*, 98 (11 January 1917), pp. 373–374. Haddon regarded Tylor as the foremost exponent of the comparative method in Britain, an evolutionist who nevertheless "was fully alive to the borrowing of culture and to cultural drifts; . . . ever since 1874 he repeatedly drew attention to the direct cultural influence of Asia on the higher civilisations of the New World". Although often critical of Tylor there seems little doubt that ES was influenced by his erudition, powers of exposition and wide-ranging speculation.

12 Dawson, p. 79.

13 W. J. Perry, *The Children of the Sun: A Study in the Early History of Civilization* (London: Methuen, 1923), p. viii. On Rivers he added: " . . . for ten years I had the benefit of his unceasing advice and sympathy, as well as, what is equally important, of his unsparing and unerring criticism". Rivers died before seeing the manuscript.

14 Perry had argued this earlier in: "The Peaceable Habits of Primitive Communities: An Anthropological Study of the Golden Age", *Hibbert Journal*, 16 (1917), pp. 28–46; "An Ethnological Study of Warfare", *Manchester Literary and Philosophical Society Memoirs*, 61 (1917), pp. 1–16. Perry also admired Buddhism for its pacifism and equality between the sexes.

15 See W. H. R. Rivers, *History and Ethnology* (London, 1922).
16 For example, Azar Gat, *War in Human Civilization* (Oxford, Oxford University Press, 2006), pp. 11–36 and *passim*.
17 Malinowski, "New and Old Anthropology", *Nature*, 113 (1 March 1924), pp. 299–301. He added: "Mr. Perry's book is the first systematic outline of a big and daring theory of human culture, and the first scheme of its birth, history, spread and partial decay".
18 See V. G. Childe, review of Perry's *Origin of Magic and Religion* (1923) and *Growth of Civilisation* (1924) in *Man*, 25 (February 1925), pp. 27–29. Perry's books, he said, "cover an enormous field of research and present a wide and comprehensive view of the whole evolution of spiritual and material civilisation . . . The result is a system which is not only calculated to appeal to the layman, but must be stimulating and suggestive to the specialist". However he was critical of some of Perry's findings in his own area, early Europe, and commented that Perry's synthesis "would be more convincing if the facts on which [he] relies had been more carefully verified, if the unifying principles had been applied in a more scientific manner, and if the omissions had been less one-sided".
19 A. C. Haddon, "The Cultural History of the Pacific", *New Zealand Journal of Science Technology*, 7 (1924), pp. 101–106.
20 O. G. S. Crawford, "The Origins of Civilisation", *Edinburgh Review*, 240 (January 1924), pp. 101, 116 (quote p. 115). He was reviewing a 1921 monograph by Perry on megalithic monuments and their distribution in England and Wales, as well as ES's new edition of *Ancient Egyptians*. Perry and Crawford continued their debate in a following issue of *Edinburgh Review*, 240 (April 1924), pp. 405–408.
21 T. Eric Peet, review of Perry, *The Origin of Magic and Religion*, in *Journal of Egyptian Archaeology*, 10 (April 1924), pp. 63–69 (quote p. 69). Peet was lecturer in Egyptology at Manchester University. He was an expert in the Italian Bronze age, Egyptian history and archaeology (especially Egyptian script). His recent works included *The Mayer Papyri* (1920) and *Egypt and the Old Testament* (1922). He had thrown doubt on any Egyptian sources for rock-hewn tombs in the megalithic areas of Europe in his *Rough Stone Monuments and Their Builders* (London, 1912), pp. 156–157. A. C. Haddon disputed some of Perry's claims about pearls in respect of Persia and Oceania in a learned paper: "Pearls as 'Givers of Life'", *Man*, 24 (December 1924), pp. 177–185. Perry replied *ibid.*, 25 (March 1925), pp. 37–41. Haddon's response criticized Perry's loose documentation but protested that he was not attacking the general theory advanced by Perry and ES: "I agree that higher cultures have entered into the Pacific and have been subject to modification and degeneration": *ibid.*, 25 (April 1925), pp. 51–53.
22 C. L. Riley *et al.*, eds, *Man Across the Sea* (Austin, London: University of Texas, 1971), pp. xiii–xiv.
23 Robert H. Lowie, review of *Children of the Sun* in *American Anthropologist*, 26 (1924), pp. 86–90 (quote p. 88). Perry propagated his ideas at the Pan-Pacific Science Congress held in Melbourne and Sydney in August-September 1923: see *Man*, 24 (January 1924), p. 13. This

congress recommended teaching and research in anthropology in Australian universities and ES was soon involved in this project.

24 Lowie, "Social Anthropology" in *Encyclopedia Britannica* (1926), pp. 566–570.

25 Lowie, *The History of Ethnological Theory* (London: Harrap, 1937). Lowie accused ES of being unscientific and referred to his "crass ignorance" in ethnography (pp. 60–61).

26 J. L. Myres, "The Historical Method in Ethnology", *The Geographical Teacher*, 13 (1925), pp. 8–28. This was the presidential address for 1925. Although not strictly a review, much of the article dealt with Perry's book.

27 W. J. Perry, "Professor Myres and The Historical Method", *ibid.*, pp. 152–154.

28 J. L. Myres, "Reply to Mr. Perry", *ibid.*, pp. 155–157. Myres made another attack on ES and Perry in his presidential address to the Folklore Society: "The Methods and Magic of Science", *Folklore*, 36 (31 March 1925), pp. 15–47. Myres said of Perry's *Children of the Sun* "that the striking and original conclusions which he claims to establish are reached by a method more akin to that of the magician than to scientific reasoning" (p. 41).

29 See my book *Darwin's Coat-Tails*, essay 13.

30 ES, "The Galton Lecture", *Eugenics Review*, 16 (1925), pp. 1–8 (quote p. 3). He gave the lecture on 18 February 1925. ES was traditional enough to assert that the Australian aborigines were of such lowly status and poorly equipped intellectually that were only capable of using a very few aspects of ancient culture, a wave of which came to them early on. Marriage regulations were said to be one of these. ES, like Perry, was convinced that the "old" theory of independent invention (sponsored by Tylor and others) was "now at its last ebb" (p. 7).

31 See my *Darwin's Coat-Tails*, essay 15, which explores the connection between American and Nazi eugenics.

32 George W. Stocking, Jr., *The Ethnographer's Magic and Other Essays in the History of Anthropology* (Madison: University of Wisconsin Press, 1992), p. 187. See Stocking for detail on this matter. He refers to the American Eugenics Research Association as the Galton Society.

33 Dawson, p. 85. Dawson details how ES rescued the chair during the Depression year of 1932 by appealing to the Prime Minister, gaining a five year extension. According to ES the term "Human Biology" was coined in discussions he had with Embree and the director of the Museum of Honolulu when he returned to New York to report on his Australian mission.

34 Dawson, p. 82.

35 ES, "Pre-Columbian Representations of the Elephant in America", *Nature* (25 November, 1915), p. 340; (16 December 1915), p. 425; (27 January 1916), p. 593.

36 See Michael D. Coe, *Breaking the Maya Code* (London: Thames and Hudson, 1992).

37 ES, *Elephants and Ethnologists* (London: Kegan Paul, 1924; New York: E. P. Dutton, 1924), p. 5. Following references in text are to this source.

ES was not the first to detect an Asian influence in the stela. The English travellers and amateur ethnologists Channing Arnold and Frederick Frost had studied the site and argued the case "that America's first architects were Buddhist immigrants from Java and Indo-China". They dated Copan as erected around the eighth century. Regarding the stela they described the dress, the ornamentation, the turban-shaped head-dress as "all purely ancient Indo-Chinese". They also considered the carvings "as so strikingly Oriental that one cannot doubt their origin. The faces of the figures on the Stelae are the faces that one can see to-day in Cambodia and Siam": Channing Arnold and Frederick J. Tabor Frost, *The American Egypt: A Record of Travel in Yucatan* (London: Hutchinson, 1909), pp. viii, 284. However in 1920 the American ethnologist S. G. Morley – in his book *The Inscriptions at Copan* – dismissed Arnold and Frost's claim as an extravagant hypothesis and accused ES of reviving this highly improbable identification. Morley was supported by other American ethnologists (eg, Tozzer, Spinden and Means). ES felt obliged to reply.

38 See previous note. ES argued that there were many features of the intricate sculpture on the Copan stelae that revealed "not only Indian but also Indo-Chinese, Indonesian and Melanesian influence – the ear-plugs and pendants, the bracelets, the anklets, the form of the girdle (and the Conus shells, so distinctive of Oceania) and many of the arbitrary forms in the ornamentation" (p. 33).

39 ES cited the extensive research of the German scholar Georg Friederici, *Malaio- Polynesische Wanderungen* (1914). This argued that the movement of seafarers from India into Oceania began very early. Polynesian travel reached a zenith between 700 and 1200 AD (which, in ES's view, coincided with the culmination of civilisation in both Cambodia and America). Some of their boats were of great size, enabling them to carry up to 200 passengers and ample provisions.

40 Anon, "Elephants in American Art", *Nature*, 114 (27 December 1924), pp. 923–925. He accused ES of "dogmatic assertion" and judged that ES's case would have gained strength "had he urged it less aggressively and devoted himself to the elimination of inconsistencies and to the more careful analysis of the evidence upon which he relies".

41 Some academics were more accepting of the ES-Perry position. One such was W. E. Armstrong, Rivers's successor in social anthropology at Cambridge (the first person to hold such a post officially). In his unpublished lectures of around 1924–25, Armstrong noted that ES's theory required both simple and cultural migration, and he asked for more evidence on the complexity of cultural contact. He commented: "The alternative to direct cultural migrations from Egypt to America is the supposition that at certain points on the heliolithic track, e.g. in India and Indonesia, interaction produced new centres of Egyptian culture which, either by the inhabitants acting as middle-men for traders from Egypt, or for their own needs, led to trading expeditions of converted aborigines (or half-castes in the beginning) in search of gold, cowries, etc.": extract from lectures in James Urry, "W. E. Armstrong and Social

Anthropology at Cambridge, 1922–1926", *Man*, new ser. 20 (September 1985), p. 424.

42 See Alice Beck Kehoe, *Controversies in Archaeology* (Walnut Creek, California: Left Coast Press, 2008), ch. 7 – further discussion in the *Afterword*).

5 The Evolution of Man (1924–1927)

1 *The Evolution of Man: Essays by G. Elliot Smith* (Oxford: Oxford University Press, 1924), p. 13. Editions were also published in London, Edinburgh, New York, Melbourne, Copenhagen, Bombay and other places. Printed in June 1924 it was reprinted in November 1924 and in later editions. References in the text are to this 1924 edition.

2 Reprinted in W. H. R. Rivers, *Psychology and Ethnology* (London, New York: Kegan Paul, 1926), pp. 120–140. This posthumous collection of papers was edited by Elliot Smith in order to make available to a wider audience the 1911 address and many other papers that were not readily accessible. In his introduction Elliot Smith wrote of the 1911 address that "it marked an epoch not only in Dr Rivers's own career, but also in the history of ethnology" (p. xxvi).

3 Darryl Forde, "Dr. W. J. Perry [1887–1949]", *Nature* (4 June 1949), pp. 865–866. Another friend Clayton Joel said: "Perry's was a happy and friendly character, blended of humour, forthrightness and keen perception, and impatient of cant and humbug. Despite the violence of the controversies in which he was involved over the Diffusion issue, he maintained the most friendly relations with most of those who publicly disputed his views": *Man*, 50 (January 1950), pp. 6–7.

4 G. W. Stocking, *The Ethnographer's Magic* (1992), pp. 190–191 and *passim*.

5 Donald Fisher, "Rockefeller Philanthropy and the Rise of Social Anthropology", *Anthropology Today*, 2 (February 1986), p. 6.

6 ES to R. M. Pearce (Rockefeller Foundation Archives), 30 October 1929, quoted Henrika Kuklick, *The Savage Within: The Social History of British Anthropology, 1885–1945* (Cambridge: Cambridge University Press, 1991), p. 211. Kuklick writes that Rockefeller "appreciated Elliot Smith's considerable skills as an anatomist, but that it would not give him additional funds for anthropological research. It had come to believe that the only truly scientific anthropology was the fieldwork-based functionalism of Malinowski's school" (p. 211).

7 ES's school at University College was initially more successful in producing postgraduates. Among anthropologists who gained a PhD there (in 1927) was A. P. Elkin, who became an eminent figure in Australian anthropology. However by the 1930s LSE was well ahead. Whereas University College produced twelve PhDs in the whole school of anatomy (which included anthropology) by1945, the anthropology school at LSE under Seligman and Malinowski produced twenty-four in the same period. LSE got more people into academic jobs in anthropology and attracted much more funding for field research: Kuklick, *Savage Within*, pp. 54–56.

8 Stocking, *Ethnographer's Magic*, p. 356. See also his *After Tylor: British Social Anthropology, 1881–1951* (London: Athlone, 1996).
9 E. E. Evans-Pritchard, "Recollections and Reflections", *The New Diffusionist*, 1 (1970), pp. 37–38. He added: "I knew, though in a rather humble student way, Elliot Smith personally. He was very kind to me, in spite of my being a student of the L.S.E., and it was he who more or less appointed me to the chair of Sociology at the Egyptian University Cairo". He also liked and admired Perry, and the two men collaborated at University College, where Evans-Pritchard was made an honorary research assistant. He later organized the appointment of Arthur Hocart, a disciple of Rivers, as his successor in the Cairo chair, "thus paying back, as it were, my debt to Elliot Smith" (p. 38).
10 W. H. R. Rivers, *Psychology and Ethnology*, edited with preface and introduction by ES (London: Kegan Paul; New York: Harcourt, Brace, 1926).
11 Brenda Z. Seligman, review in *British Journal of Psychology*, 17(1926–1927), pp. 370–376. The criticism was not quite fair, for, as we have seen, diffusionists such as ES and Perry accepted the evolution of higher civilizations in India and China, while claiming that Egypt also contributed elements to those cultures.
12 ES, "Prehistoric India", letter to *Times* (8 March 1926), p. 10.
13 ES, "Rock Carvings in New Zealand", letter to *Times* (15 March 1926), p. 10.
14 Raymond Firth, "Maori and Egyptian", letter to *Times* (23 March, 1926), p. 12.
15 *Times*, 21 August; 2, 7, 8, 9, 17, 18, 23, 27 September; 16 October, 1926.
16 J. L Myres, "Myth and History", letter to *Times* (8 September 1926), p. 13; ES, *ibid*., (9 September 1926), p. 8.
17 J. Eric Thompson, "The Elephant Heads in the Waldeck Manuscripts", *The Scientific Monthly*, 25 (November 1927), p. 392.
18 ES, "Elephants or Macaws? Asia and American Civilization", *Times* (14 January 1927), pp. 13, 16. See also his article "The 'Elephant Controversy' Settled by a Decisive Discovery", *Illustrated London News* (15 January 1927), pp. 86, 108. ES took biographical information on Waldeck (cited above) from the American historian Hubert Howe Bancroft.
19 H. J. Braunholtz, "Elephants or Macaws", letter to *Times* (22 January 1927), p. 8. He also claimed that Waldeck's representations of some of the glyphs from the "Temple of Inscriptions" at Palenque produced realistic elephant's heads "out of the most unpromising material". Braunholtz was to repeat his criticisms in *Man* in 1958, long after ES's death. He added a quotation from a French writer, C. J. D. Charnay in 1885, noting Waldeck's penchant for elephants: Braunholtz, letter to *Man*, 124 (June 1958), pp. 95–96. Braunholtz (1888–1963) became Keeper of the Ethnological Department in the British Museum from 1945–1953: "Obituary", *Times* (6 June 1963), p. 17.
20 A. P. Maudslay, "The Maya Sculptures", letter to *Times* (14 February 1927), p. 8.
21 ES, letter to *Times* (25 January 1927), p. 8.

22 On both Kingsborough and the Israel myths see Robert Wauchope, *Lost Tribes and Sunken Continents* (Chicago: University of Chicago Press, 1962), especially ch.4. A young Irish Viscount, Kingsborough went bankrupt publishing "a magnificent set of nine imperial folio volumes reproducing Mexican codices and commenting on Mexican antiquities" (p. 52). He died in debtor's prison aged forty-two.

23 Wikipedia (entry on Waldeck), and generally; see also Howard F. Kline, "The Apocryphal Early Career of J. F. de Waldeck, Pioneer Americanist", *Acta Americana*, 5 (1947), pp. 278–300; and Michael D. Coe, *Breaking the Maya Code* (1992), pp. 76–77. Coe cites experts who assess Waldeck's original working drawings in the Ayer Collection as being of high quality: "In spite of that, one can put no credence in his finished lithographs, which have always been treated with disdain by Mayanists, and rightly so" (p. 77).

24 Thompson, "Elephant Heads", pp. 392–398.

25 ES, "The Philosophical Background of Ethnological Theory", *Journal of Philosophical Studies*, 2 (1927), pp. 182–189.

26 ES, "Some Aspects of Rationalism", *The Rationalist Annual* (1927), pp. 19–24 (quote p. 24). He ended on a cynical note: "No one who has been involved in controversies relating to the problems of the Humanities can fail to be oblivious of the fact that pure reason carries little weight in argument when it comes into competition with tradition and traditional catch-phrases" (p. 24). He had made a similar point a few years earlier. In 1925, during the celebrated Scopes trial, when a Tennessee high school teacher was prosecuted for teaching evolution in defiance of state law, ES joined other English scientists in protest. He declared that "the Tennessee comedy" was "essentially the three-century-old attempt to destroy intellectual freedom . . . But the reality of evolution is as certain as the fact that the earth revolves around the sun": *Nature*, 2906 (11 July 1925), p. 75.

27 ES, "The Idea of the Supernatural in Human Development", *Encyclopedia of Modern Knowledge*, 1 (1936), p. 375.

28 ES, *Human Nature* (London: Watts, 1927), pp. 16–17. He made this point, he said, "lest you imagine I am going to commit myself to the pessimistic doctrine of assuming progress to be attainable only by breeding and not by education" (p. 17). The lecture was given on 17 March 1927.

29 R. R. Marett, "The Diffusion of Culture", the Frazer Lecture for 1927; reprinted in Warren Dawson, ed., *The Frazer Lectures, 1922–1932* (London: Macmillan, 1932), pp. 172–189.

30 R. R. Marett, *A Jerseyman at Oxford* (Oxford University Press, 1941).

31 Marett, "Diffusion of Culture", p. 175. Following quotes are from this source.

32 Roland B. Dixon, *The Building of Cultures* (New York: Scribner, 1928).

6 The Diffusion Controversy (1927–1933)

1 B. Malinowski "The Life of Culture", *Forum*, 76 (1926), reprinted in ES *et al.*, *Culture: The Diffusion Controversy* (see n2), p. 27.

2 "Is Civilization Contagious?" *Forum*, 76 (1926), pp. 171ff. ES argued the "yes" case ("The Diffusion of Culture") and Malinowski the "no" case ("The Life of Culture").

3 G. Elliot Smith, B. Malinowski, Herbert J. Spinden, Alex. Goldenweiser, *Culture: The Diffusion Controversy* (New York: W. W. Norton, 1927). The essays were as follows: ES, "The Diffusion of Culture", pp. 9–25; Malinowski, "The Life of Culture", pp. 26–46; Spinden "The Prosaic Vs. The Romantic School in Anthropology", pp. 47–98; Goldenweiser, "The Diffusion Controversy", pp. 99–106.

4 For more on Spinden see Robert L. Brunhouse, *Pursuit of the Ancient Maya: Some Archaeologists of Yesterday* (Albuquerque: University of New Mexico Press, 1975), chapter 5.

5 ES to C. M. Hincks (January 1926), quoted Dawson, p. 93 (letter in full, pp. 89–95). ES admitted the inspiration he had got from Rivers's example. Rivers's broad outlook stemmed from his personal experience as a clinical physician, as neurologist, expert on the physiology of sense organs, as experimental psychologist, anthropologist and clinician in psychiatry (p. 91).

6 Donald Johanson and James Shreeve, *Lucy's Child: The Discovery of a Human Ancestor* (London: Viking, 1989), p. 58. Dart appeared "with only the Taung skull cupped in his hands, and with very little to say that the audience had not already read in print". The authors argue that Dart thus lost his opportunity to impress his views on people, that the Taung skulll "faded out of the public eye" and that Dart's findings were unfairly neglected for many years. ES was in fact generally a supporter of his student Dart's findings. However as Zuckerman pointed out ES did judge a 1931 paper by Dart on *Austalopithecus* as "falling well below the standards of a scientific contribution": *Concepts of Human Evolution*, p. 12. See Dawson for information on ES's activities at this time.

7 As his student A. P. Elkin recalled, on one occasion "he told us how nonplussed he was on the occasion of his first radio broadcast. He was a gifted public lecturer, needing no notes, so when the radio medium became available he agreed to use it. So into the studio he went, as on the platform, but seeing no one to talk to, only a metal gadget, he was momentarily struck dumb. However he managed to get started and then all was all right. But, after that, he always took some notes so that he could get going": A. P. Elkin, "A Personal Testimony" in A. P. Elkin and N. W. G. Mackintosh, eds, *Grafton Elliot Smith: The Man and His Work* (Sydney: Sydney University Press, 1974), p. 12.

8 ES, *Human History* (London: Cape, 1930), p. 497. H. A. L. Fisher's writings on the glorious Greek legacy made a strong impact on ES (p. 460). ES was also influenced by P. N. Ure's *The Origin of Tyranny* (1922). Ure traced the beginnings of revolt against the whole Egyptian despotic state system to the Ionian invention of metal currency, "perhaps the most

epoch-making revolution in the whole history of commerce", which ushered in a more individualistic and questioning mode of life (Ure, quoted, p. 424).

9 *Survey*, Vol. 69 (February 1930), p. 598.

10 *The Nation* commended its attempts to unify biological and cultural history: 130 (4 June 1930), p. 656 (by William MacDonald), while the *Times Literary Supplement* pronounced it a "masterly synthesis, even if (one's) immediate reaction to it is to try to pull it to pieces: (3 April 1930), p. 287. *The Saturday Review* thought it "interesting and readable; but it is overloaded with guesses": 149 (1 March 1930), p. 269, while *The New Statesman* declared: "The underlying principle of the whole outlook strikes us as unsound. Why is it a law of anthropology that connections must be found at all costs between all cultures, however widely differing in nature or geographical position?": 34 (8 March 1930), p. 714. *The Nation and Athenaeum* said: "Diffusion as a working hypothesis of human progress is a boon to anthropology; as a dogma it promises to become a nuisance": 47 (26 April 1930), p. 117. The sociologist Alexander Farquharson recognised ES's "great capacity and ingenuity; perhaps one might add – intrepidity" and sided with him on the diffusion of culture elements throughout the world from their centres of origin. However (as one sympathetic to the Geddes-Le Play school) Farquharson judged ES's almost total scepticism about "the geographical conditioning of human affairs" to be naïve, even partisan: *Sociological Review*, 23 (1931), p. 32. Margaret Mead described the book as "a bulky and fanciful excursion into the unknown past by a famous anatomist who believes that the invention and spread of civilization has been the ruin of the human race": brief notice in *The Annals of the American Academy*, 152 (Nov. 1930), p. 422.

11 *Early Man: His Origin, Development and Culture* by G. Elliot Smith, Sir Arthur Keith, F. G. Parsons, M. C. Burkitt, Harold J. Peake and J. L. Myres (London: Benn, 1931), ES's essay pp. 13–46.

12 ES, *The Search for Man's Ancestors* (London: Watts, 1931).

13 W. Le Gros Clark in *Man* (November 1932), p. 263. Le Gros Clark thought that the last chapter "provides a particularly illuminating discussion of the location of the birthplace of mankind".

14 ES, *In the Beginning*, new and revised edition (London: Watts, 1932), 109pp, price one shilling. For a generally favourable review by the Egyptologist G. D. Hornblower see *Man* (December 1932), pp. 285–286.

15 ES, *The Diffusion of Culture* (London: Watts, 1933), quoted p. 2. Following quotes in text are from this source.

16 Robertson in fact wavered between saying that there were real resemblances, or casual resemblances between American peoples and others, and saying that their customs differed remarkably from other humans. ES did not fail to underline these inconsistencies.

17 On this see Marvin Harris, *The Rise of Anthropological Theory* (New York: Columbia, 1968), pp. 174–175. Harris says: "That Tylor regarded independent inventions as a strong argument for psychic unity does not mean that he regarded diffusion as a strong argument *against* psychic unity" (p. 175).

18 As O'Leary had aptly put it, Mohammedanism provides us with "the most romantic history of culture drift which is known to us in detail": quoted p. 34.

7 Last Days (1933–1937)

1 Dawson, pp. 102–103 (quote above from a letter ES to Dawson, 2 November 1931, p. 100).
2 S. Zuckerman, "Sir Grafton Elliot Smith: 1871–1937" in Professor Lord Zuckerman, ed., *The Concepts of Human Evolution* (symposium in honour of Elliot Smith, 9–10 November 1972, published by the Zoological Society of London by Academic Press, London, 1973), p. 19. Zuckerman added that H. A Harris, in charge during ES's illness, "was vigorous, ambitious and possessive, and not the most popular member of the staff of the Department. He refused to allow anyone to visit Elliot Smith in hospital. One day, however, when I was about to leave to take up an appointment at Yale, I insisted on seeing him to say goodbye. Elliot Smith had been prepared for my visit, and was sitting up in bed with a cheroot stuck between his twisted lips. His opening words, in a voice which I hardly recognized, were "Why have you kept away?'. I didn't dare give him the reason" (p. 19).
3 H. A. Harris, "At University College, London" in Dawson, p. 182.
4 "British Anthropologist Face the Nazis", essay 13 in Paul Crook, *Darwin's Coat-Tails* (Oxford, New York: Peter Lang, 2007). This essay uses ES's papers at the British Library [BL], Add MSS 56303-004. There are two volumes of autobiographical fragments and letters, etc. [hereafter ESP].
5 The minutes of the founding meeting of the Human Biology Research Committee give the full text of ES's introductory remarks on purposes, scale and procedure. They are to be found in RAI Archives, London, Book A23 (4 March 1932), reprinted in *Man*, 32 (April 1932), p. 95.
6 ES to C. G. Seligman (5 March 1932), ESP Add MSS 56303, pp. 514–515 (BL).
7 ES, "Notes for Discussion of Purposes, Scope and Procedures of the Human Biology Committee" in ESP, Add MSS 56303, p. 512 (BL).
8 ES to *Times*, 22 May 1934, p. 15.
9 Marginal note by Myres on clipping from *Times*, 22 May 1934 in Myres papers, Box 121, fol.127 (Bodleian Library, Oxford).
10 A. A. Abbie (University of Adelaide), "Sir Grafton Elliot Smith" in *Bulletin of the Post-Graduate Committee in Medicine, University of Sydney*, 15, 3 (1959), pp. 131–132.
11 ES, "Chairman's Address" to section Aa in *Congrès International* (1934), pp. 65–68; also *Times*, August 1934, p. 9.
12 *New York Times*, 1 August 1934, p. 19.
13 ES to Seligman, 15 August 1934: ESP, Add MSS 56303, pp. 527–528.
14 ES to Boas, 8 October 1934 in Boas Papers, Philadelphia, 71. Boas's pamphlet was widely distributed in Europe in 1934, circulating underground in Germany.

15 ES, "The Aryan Question", *Rationalist Annual* (1935), pp. 30–34.
16 He expressed these views in a letter to Firth (16 June 1935), Myres Papers (Oxford), Box 121, fol 238.
17 For a discussion of Julian Huxley's role in events at this time see essays 14 and 18 in my book *Darwin's Coat-Tails*. His book was co-authored with A. C. Haddon, but Huxley wrote most of it.
18 *Race and Culture* (London: Le Play House Press, nd [1936], price one shilling), pp. 5–6.
19 G. M. Morant, "Racial Theories and International Relations", *Journal of the Royal Anthropological Institute*, 69 (1939), pp. 151–162.
20 ES, "The Place of Thomas Henry Huxley in Anthropology", *Journal of the Royal Anthropological Institute*, 65 (1935), pp. 199–204.
21 Dawson, p. 107. In a letter to Dawson on 2 November 1936, ES wrote: "Just after you called on us last at Albert Road, my wife had the misfortune to be involved in an explosion of a gas stove and for the burns to the face my son promptly had her sent to U. C. Hospital and decided it would be a good thing to park both his parents during the moving" (p. 109). This was to Queen Mary's Hospital at Sidcup in Kent.
22 Quoted in Zuckerman, p. 3.

AFTERWORD

1 Quoted in Gerhard Kraus, "G. Elliot Smith (and W. J. Perry) on Trial", *New Diffusionist*, 4 (1974), p. 64.
2 Glyn Daniel, *The Idea of Prehistory* (Pelican edition, 1964), pp. 95, 97, 98 and *passim*.
3 Colin Renfrew, "Daniel, Glyn Edmund (1914–1986)", rev. *Oxford Dictionary of National Biography* (Oxford University Press, 2004) <*http://www.oxforddnb.com/view/article/39803*>; Glyn Daniel, *Some Small Harvest: The Memoirs of Glyn Daniel* (London: Thames and Hudson, 1986).
4 Glyn Daniel, *The First Civilizations: The Archaeology of Their Origins* (London: Thames and Hudson, 1968), pp. 191–192. The seven civilizations were Mesopotamia, Egypt, Olmec-Zapotec, Maya, Peru, Indus and Yellow River.
5 *Ibid*., pp. 189–190; Betty J. Meggers, Clifford Evans and Emilio Estrada, *The Early Formative Period on Coastal Ecuador* (Washington: Smithsonian, 1965).
6 *The First Civilizations*, pp. 186–187.
7 Professor Lord Zuckerman, ed., *The Concepts of Human Evolution* (London, New York: Academic Press, 1973) [hereafter Zuckerman].
8 According to Gerhard Kraus: "To balance the predictably adverse impact of a Glyn Daniel, we suggested to the Zoological Society that they allowed us to present the case for Elliot Smith as Diffusionist. This was rejected, but it was agreed that we would be given an opportunity to put the case from the floor during discussion time. In the event, the official speakers and the chairman occupied the floor for almost all the time available with their mainly anti-Elliot Smith views, while we were left with

only ten minutes or so for making our points": Kraus, "Elliot Smith on Trial", pp. 62–63. There may be a hint of conspiracy theory here. To be fair, three diffusionists had their comments published in the proceedings (seven pages in all).

9 Daniel, "Elliot Smith, Egypt and Diffusionism" in Zuckerman, pp. 421, 414 (Daniel's paper is pp. 407–424; subsequent discussion pp. 424–447). Following references in the text are to Zuckerman.

10 When he could find no transitional type between the serdab of Egypt (a chamber found in some mastabas containing a statue of the tomb-owner) and the European dolmen he "invented a transitional form of which he drew a section and plan . . . This was a purely hypothetical monument but Elliot Smith remained unperturbed". He was guessing at the rough form that it must have assumed: "Trait-chasing was one thing, making false analogies was another, but inventing transitional forms was not merely a negation of method but positively wicked" (p. 417).

11 C. L. Riley, J. C. Kelley, C. W. Pennington and R. L. Rands, eds, *Man Across the Sea: Problems of Pre-Columbian Contacts* (Austin: Texas University Press, 1971: hereafter *Man Across the Sea*, discussed below). Elsewhere Kraus praised the work of the German scholar Wolfgang Marschall, who in *Transpazifische Kulturbeziehungen* (Munich: Renner, 1972) "shows that American pre-Columbian, trans-Pacific culture contacts were both numerous and irrefutable": Kraus, review in *Man*, 8 (March 1973), pp. 115–116.Marschall illustrated his case from four selected traits: the blowpipe, house-models in connection with burials, animal figurines on wheels, and weaving and dyeing techniques. Gerhard Kraus had published his book *Homo Sapiens in Decline: A Reappraisal of Natural Selection* in 1973.

12 Daniel, in his *Megalithic Builders of Western Europe* (Harmondsworth: Pelican, 1963, p. 25), had pictured ES and Perry spending a lifetime "advocating that civilization was due to the spread of a master-race from Egypt", while E. Leach at the symposium alleged a racial basis behind ES's theory (pp. 432ff). Kraus found it necessary in his writings to refute such views. He pointed out, correctly, that ES and Perry had been critical of the Pharaoh system as "one of the major aberrations that affected mankind adversely and that its world-wide consequences (i.e. dynastic rule and the consequent warfare among sovereign states) may have led us into our present predicament. This is certainly not the way in which one would speak of a 'Master Race' or about the 'Saviours' of the world". He also pointed out that Daniel had drastically simplified ES's views when he accused him of saying that the Egyptians had *directly* hawked their civilization all around the world: Kraus, "ES on Trial", pp. 79–80.

13 *Guardian Weekly*, 26 November 2010.

14 G. F. Ekholm, "The Possible Chinese Origin of Teotihuacán Cylindrical Tripod Pottery", *Proceedings of the Thirty-Fifth International Congress of Americanists* (1964), pp. 39–45; R. Heine-Geldern, "Chinese Influence in the Pottery of Mexico, Central America, and Columbia", *Proceedings of the Thirty-Third International Congress of Americanists* (1959), pp. 207–210; Ekholm and Heine-Geldern, "Significant Parallels in the

Symbolic Arts of Southern Asia and Middle America" in S. Tax, ed., *The Civilizations of Ancient America* (Chicago, 1959), 1: pp. 299–309. I am indebted to Kehoe (next note) for this information.

15 On this and some of the factors mentioned above see Kehoe (discussed below).

16 See, for example, my essay "Eugenics, Genetics and Feeblemindedness" in Paul Crook, *Darwin's Coat-Tails* (essay 16).

17 John L. Sorenson, "The Significance of an Apparent Relationship between the Ancient Near East and Mesoamerica" in *Man Across the Sea* (1971), pp. 220, 224. 225. Sorenson dismissed early diffusionism as naïve. However an anthropology that "fails to control for historical diffusionism seems to rest on grounds potentially as weak as unrestrained diffusionism which lacks appreciation for the adaptive power of man" (p. 226). Sorenson felt that the best scholarship "appears to have demonstrated that civilization – that most complex manifestation of man's cultural behavior – wherever it manifests itself in the Old World, is part of one great ecumenical web. If it now turns out that New World civilization is seriously related to (that is, dependent upon) that of the Old World, then there can be no ultimately satisfying scientific answer to the question of how men became civilized, for there will be but a single case of the phenomenon. One cannot generalize from a single case" (p. 225).

18 *Man Across the Sea*, p. 448. As the editors add: "It must be stressed that these similarities represent, not actual spread of artifacts, but possible copying of one technique or design element of one culture by another. If actually copied, these examples obviously are proof of contact from one area to the other, but, of course, the question here centers on the meaning of the resemblances" (p. 448).

19 *Ibid.*, pp. 443, 446, 454, 457–458. For an interesting discussion that places diffusionism within the context of western colonialism and Eurocentric values in historical and geographical scholarship see J. M. Blaut, *The Colonizer's Model of the World* (New York, London: Guilford Press, 1993).

20 Alice Beck Kehoe, *Controversies in Archaeology* (Walnut Creek, California: Left Coast Press, 2008). Citations for the scholars mentioned above may be found in Kehoe. Page numbers in the text refer to Kehoe's book.

21 It should be said that she holds no brief for ES, seeming to consign him (unfairly in my view) to the box of misguided or eccentric enthusiasts.

22 "Geographer Carl Johannessen has compiled a list of more than two hundred organisms, both plant and animal (including insects and disease pathogens), evidencing pre-Columbian transoceanic contacts between the Americas and Eurasia or Africa." (She also notes that cocaine, which comes only from the South American coca plant, "has been identified in Egyptian mummies of the first millennium BCE by a forensic chemist"). "Most archaeologists don't want to think about this evidence; it makes them uncomfortable to discuss data drawn from research areas outside their own studies, and it's hard to overturn the reigning paradigm that 'primitive people' couldn't cross water 'barriers' . . . Students wanting to study transoceanic diffusion are told by their professors to simply avoid dealing with any proponents of pre-Columbian ocean crossings . . . It

usually happens that researchers with good scholarly credentials and mainstream work find that their 'thinking out of the box' is politely ignored. This is largely the result of the culture of the university . . . There is also legitimate worry that unorthodox opinions may jeopardize getting a job; the older generation who hires younger researchers and teachers may not welcome challenges to their ideas and procedures" (pp. 159, 160,169).

Select Bibliography

Elliot Smith papers are deposited in the British Library; University College, London University; Medical Research Council, London; John Rylands Library, Manchester; Royal Anthropological Institute of Great Britain and Ireland; National Library of Scotland; Australian Academy of Science, Canberra; and Cambridge University Library.

Extracts from his letters are conveniently printed in Warren Dawson, "A General Biography" in Warren R. Dawson, ed., *Sir Grafton Elliot Smith: A Biographical Record By His Colleagues* (London: Jonathan Cape, 1938), pp. 17–112 [hereafter Dawson]. Dawson, a student and colleague of Elliot Smith, had access to much of his correspondence and other sources, some of which are now lost or unavailable. Dawson's sketch is to date the only available biographical study of Elliot Smith. There is much useful information in other commemorative volumes: Professor Lord Zuckerman, ed., *The Concepts of Human Evolution* (London: Academic Press, 1973), proceedings of a symposium in honour of Elliot Smith organized by the British Anatomical Society and the Zoological Society of London, held in 1972; and A. P. Elkin and N. W. G. Mackintosh, eds, *Grafton Elliot Smith: The Man and His Work* (Sydney: Sydney University Press, 1974).

A detailed bibliography of Elliot Smith's writings is to be found in Dawson, pp. 219–253. There are 434 items, many specialist and technical papers. Below is given a selection of his main writings that are of relevance to this study. It is followed by a selection of some other contemporary writings and secondary commentaries that may be of use to readers who wish to follow up themes discussed in this book.

Main Writings of Elliot Smith

The Ancient Egyptians and their Influence upon the Civilization of Europe (London, New York: Harper, 1911). (The second edition of 1923 was changed to *The Ancient Egyptians and the Origin of Civilization.*)

"The Evolution of the Rock-cut Tomb and the Dolmen" in *Essays and Studies Presented to William Ridgeway* (Cambridge: Cambridge University Press, 1914), pp. 493–546.

The Migrations of Early Culture (Manchester: Manchester University Press; London: Longmans Green, 1915).

"Ships as Evidence of the Migrations of Early Culture", *Journal of the Manchester Egyptian and Oriental Society*, 5 (1916), pp. 63–102.

The Evolution of the Dragon (Manchester: Manchester University Press, 1919).

"Anthropology", *Encyclopaedia Britannica*, 12th edition (1921), pp. 143–154.
Tutankhamen and the Discovery of his Tomb (London, New York: Routledge, 1923).
The Evolution of Man (Oxford: Oxford University Press, 1924).
Elephants and Ethnologists (London: Kegan Paul, 1924).
Human Nature (London: Watts, 1927).
In the Beginning: The Origin of Civilization (London: Howe, 1928).
Human History (London: Cape, 1930).
The Search for Man's Ancestors (London: Watts, 1931).
The Diffusion of Culture (London: Watts, 1933).
He also edited (with B. Malinowski, H. J. Spinden and A. Goldenweiser), *Culture: The Diffusionist Controversy* (New York: N. W. Norton, 1927).

Other Commentaries

Abbie, A. A., "Sir Grafton Elliot Smith", *Bulletin of the Post-Graduate Committee in Medicine, University of Sydney*, 15,3 (1959), pp. 131–132.
Arnold, Channing and Frost, F. J. T., *The American Egypt: A Record of Travel in Yucatan* (London: Hutchinson, 1909).
Belich, James, *Replenishing the Earth: The Settler Revolution of the Anglo-world* (Oxford: Oxford University Press, 2009).
Bishop, M. J., ed., *The Cave Hunters: Biographical Sketches of the Lives of Sir William Boyd Dawkins (1836–1929) and Dr. J. Wilfrid Jackson (1880–1978* (Derbyshire Museum Service, 1982).
Blaut, J. M., *The Colonizer's Model of the World* (New York, London: Guilford Press, 1993).
Brunhouse, Robert L., *Pursuit of the Ancient Maya: Some Archaeologists of Yesterday* (Albuquerque: University of New Mexico Press, 1975).
Coe, Michael D., *Breaking the Maya Code* (London: Thames and Hudson, 1992).
Crawford, O. G. S., "The Origins of Civilization", *Edinburgh Review* 240 (January 1924), pp. 101–116.
Crook, Paul, *Darwin's Coat-Tails: Essays on Social Darwinism* (Oxford, New York: Peter Lang, 2007).
Daniel, Glyn, *Megalithic Builders of Western Europe* (Harmondsworth: Pelican, 1963); *The Idea of Prehistory* (Harmondsworth: Pelican, 1964); *The First Civilizations: The Archaeology of Their Origins* (London: Thames and Hudson, 1968); *Some Small Harvest: The Memoirs of Glyn Daniel* (London: Thames and Hudson, 1986).
Dawson, Warren R, ed., *Sir Grafton Elliot Smith: A Biographical Record By His Colleagues* (London: Jonathan Cape, 1938).
Dixon, Roland B., *The Building of Cultures* (New York: Scribner, 1928).
Drower, Margaret S., *Flinders Petrie: A Life in Archaeology* (Madison: London, University of Wisconsin Press, 1985, 2nd ed, 1995).
Elkin, A. P. and Mackintosh, N. W. G., eds, *Grafton Elliot Smith: The Man and His Work* (Sydney: Sydney University Press, 1974).
Evans-Pritchard, E. E., "Recollections and Reflections", *The New Diffusionist* 1 (1970), pp. 37–38.

Fisher, Donald, "The Rockefeller Foundation and the Development of Scientific Medicine in Great Britain", *Minerva*, 16 (1978), pp. 20–41.

Forde, C. Daryll, *Ancient Mariners* (London: Howe, 1927); "Dr. W. J. Perry", *Nature*, 163 (4 June 1949), pp. 865–866.

Gat, Azar, *War in Human Civilization* (Oxford: Oxford University Press, 2006).

Goldenweiser, Alexander. A., "Diffusion versus Independent Origin: A Rejoinder to Professor G. Elliot Smith", *Science*, ns 44 (13 October 1916), pp. 531–533; *Early Civilization: An Introduction to Anthropology* (New York: Knopf, 1922).

Jackson, J. Wilfrid, *Shells as Evidence of the Migration of Culture* (Manchester: Manchester University Press, 1917).

Joel, C. E., "Elliot Smith and the Macleay Mummy", *Man*, ns 4, 4 (December 1969), pp. 645–646; "William James Perry, 1887–1949", *Man* (January 1950), pp. 6–7.

Johanson, Donald and Shreeve, James, *Lucy's Child: The Discovery of a Human Ancestor* (Bungay, Suffolk: Viking, 1990).

Kehoe, Alice Beck, *Controversies in Archaeology* (Walnut Creek, California: Left Coast Press, 2008).

Kline, Howard F., "The Apocryphal Early Career of J. F. de Waldeck, Pioneer Americanist", *Acta Americana*, 5 (1947), pp. 278–300.

Kraus, Gerhard, "G. Elliot Smith (and W. J. Perry) on Trial", *The New Diffusionist* 4 (1974), pp. 62–109; *Homo Sapiens in Decline* (Great Gransden: New Diffusionist Press, 1973).

Kuklick, Henrika, *The Savage Within: The Social History of British Anthropology, 1885–1945* (Cambridge: Cambridge University Press, 1991).

Langham, Ian, *The Building of British Anthropology* (London: Reidel, 1981).

Lewis, A. L., "Dolmens or Cromlechs", *Man*, 11, 98 (1911), p. 175.

Lowie, Robert H., "Social Anthropology", *Encyclopedia Britannica* (1926), pp. 566–570; *The History of Ethnological Theory* (London: Harrap, 1937).

Malinowski, Bronislaw, "New and Old Anthropology", *Nature*, 113 (1 March 1924), pp. 299–301.

Marett, R. R., "The Diffusion of Culture" in Warren Dawson, ed., *The Frazer Lectures, 1922–1932* (London: Macmillan, 1932); *A Jerseyman at Oxford* (Oxford: Oxford University Press).

Means, P. A., "Some Objections to Mr. Elliot Smith's Theory", *Science*, ns 44 (13 October 1916), pp. 533–534.

Meggers, Betty J. *et al.*, *The Early Formative Period on Coastal Ecuador* (Washington: Smithsonian, 1965).

Morant, G. M., "Racial Theories and International Relations", *Journal of the Royal Anthropological Institute*, 69 (1939), pp. 151–162.

Myres, J. L., "The Historical Method in Ethnology", *The Geographical Teacher*, 13 (1925), pp. 8–28; "The Methods and Magic of Science", *Folklore*, 36 (31 March 1925), pp. 15–47.

Nuttall, Zelia, "The Fundamental Principles of Old and New World Civilizations", *Archaeological and Ethnological Papers of the Peabody Museum* (2 March 1901).

O'Leary, De Lacy, *Arabic Thought and its Place in History* (London: Kegan Paul, 1922).

Pear, T. H., "Some Early Relations between English Ethnologists and Psychologists", *Journal of Royal Anthropological Institute of Great Britain and Ireland*, 90 (1960), pp. 228–229.

Peet, T. Eric, *Rough Stone Monuments and Their Builders* (London: Harper, 1912); *The Mayer Papyri* (London: Egypt Exploration Society, 1920); *Egypt and the Old Testament* (London: Hodder and Stoughton, 1922).

Perry, W. J., "The Peaceable Habits of Primitive Communities", *Hibbert Journal*, 16 (1917), pp. 28–46; "An Ethnological Study of Warfare", *Manchester Literary and Philosophical Society Memoirs*, 61 (1917), pp. 1–16; *The Megalithic Culture of Indonesia* (Manchester: Manchester University Press, 1918); *The Children of the Sun: A Study in the Early History of Civilization* (London: Methuen, 1923); *Origin of Magic and Religion* (London: Methuen, 1923); *The Growth of Civilization* (1924); *Gods and Men* (London: Howe 1927); *The Primordial Ocean* (London: Methuen, 1935); *The Growth of Civilization* (Harmondsworth: Penguin, 1937).

Pretty, Graeme L., "The Macleay Museum Mummy from Torres Straits", *Man*, ns 4 (March 1969), pp. 27–39 .

Rivers, W. H. R., *History and Ethnology* (London: SPCK; New York: Macmillan, 1922); *Psychology and Politics and other Essays* (London: Kegan Paul, 1923); *Psychology and Ethnology* (London: Kegan Paul, 1926).

Riley, C. L. *et al.*, eds, *Man Across the Sea: Problems of Pre-Columbian Contacts* (Austin, London: University of Texas Press, 1971).

Slobodin, Richard, *W. H. R. Rivers* (New York: Columbia University Press, 1978).

Stocking, George W., Jr, *The Ethnographer's Magic and Other Essays in the History of Anthropology* (Madison: University of Wisconsin Press, 1992); *After Tyler: British Social Anthropology, 1881–1951* (London: Athlone, 1996).

Thompson, J. Eric, "The Elephant Heads in the Waldeck Manuscripts", *The Scientific Monthly*, 25 (November 1927), p. 392.

Trigger, Bruce T., *A History of Archaeological Thought* (Cambridge: Cambridge University Press, 1989, repr. 2006).

Tyler, E. B., *The Early History of Mankind* (London: Murray, 1865); *Primitive Culture* (London: Murray, 1873).

Ure, P. N., *The Origin of Tyranny* (Cambridge: Cambridge University Press, 1922).

Urry, James, "W. E. Armstrong and Social Anthropology at Cambridge, 1922–1926", *Man*, ns (September 1985), pp. 412–433; *Before Social Anthropology: Essays on the History of British Anthropology* (Camberwell, Victoria: Harwood Academic Press, 1993).

Wauchope, Robert, *Lost Tribes and Sunken Continents* (Chicago: University of Chicago Press, 1962).

Wilson, J. T., "Sir Grafton Elliot Smith", *Obituary Notices of Fellows of the Royal Society*, 2 (January 1938), pp. 323–324.

Zuckerman, Professor Lord [Solly], ed., *The Concepts of Human Evolution* (London: Academic Press, 1973).

Index